W9-CPL-759

Praise for *Wonderdog*

"*Wonderdog* is a paean to these clever, flexible, charming animals who sit and walk alongside us—and also a humane, thoughtful consideration of the science using and about dogs. You'll want to read it with a dog by your side, so you can regularly turn to them admiringly and tickle their ears."

—Alexandra Horowitz, #1 *New York Times* bestselling author of *Inside of a Dog*

"A wonderful book! I loved it. Informative and engaging."

—Virginia Morell, author of *Animal Wise*

"A fresh and vibrant account of what we've learned about dogs from Darwin to today. With a cast of familiar and almost-forgotten characters, *Wonderdog* tells us why dogs do the things they do—and what it tells us about ourselves. Full of compassion and intrigue, this is scientific storytelling at its very best."

—Zazie Todd, author of *Wag: The Science of Making Your Dog Happy*

"Zoologist Howard enlists the help of veterinary professionals, psychologists, ethologists, neurologists, historians, and others in this eclectic history of dogs. Howard peppers in charming stories of his own childhood dog, Biff, giving the survey equal parts heft and heart: 'We had all the hallmarks of love for one another, Biff and I.' This is just the thing for dog lovers."

—*Publishers Weekly*

"*Wonderdog* offers readers a whirlwind tour of one hundred and fifty years of research on the minds and behavior of man's best friend. From Darwin and Pavlov to the latest research in canine science, *Wonderdog* reflects first-rate scholarship yet reads like a detective novel. This book puts Jules Howard in the top ranks of contemporary science writers."

—Hal Herzog, author of *Some We Love, Some We Hate, Some We Eat: Why It's So Hard to Think Straight About Animals*

"*Wonderdog* is a wonderful, fact-filled, and easy-to-read journey into the heads and hearts of dogs—who they are, what they know, and what they feel. It's essential to know and respect how these fascinating animals sense their worlds so that we can help them adapt to ours—so they can get all they need as they negotiate a human-oriented world. Howard does a masterful job blending the latest science with doses of common sense as he covers what we know and still need to know to give dogs the best lives possible. *Wonderdog* is a must-read."

—Marc Bekoff, University of Colorado, author of *Canine Confidential: Why Dogs Do What They Do* and co-author with Jessica Pierce of *Unleashing Your Dog: A Field Guide to Giving Your Canine Companion the Best Life Possible*

"Turning wolves into dogs took knowledge, insight and a few cheeky treats along the way. This book contains all three, and is the perfect companion to any dog lover."

—Ben Garrod, evolutionary biologist and conservationist

"With *Wonderdog*, Jules Howard explores the highs and the lows of science's sometimes troubled relationship with the domesticated wolf with which we share our homes and lives. With his characteristic lightness of touch, Howard takes us on a journey of discovery that will leave no dog-lover unmoved, and no dog-hater unconverted. A splendid, entertaining, and hugely informative read!"

—Adam Hart, author of *Unfit for Purpose: When Human Evolution Collides with the Modern World*

"A brilliant history of how we came to know our best friends better—the trials and tribulations, the highs and lows. Jules Howard reveals how we came to know dogs better and how that's helped us understand ourselves."

—Professor Alice Roberts, biological anthropologist, broadcaster, and author of *Ancestors*

"The book about dogs I never knew I needed, full of answers to questions I never thought to ask. A fascinating and eye-opening read for anyone that has ever loved a dog."

—Jess French, veterinarian, zoologist, broadcaster, and author of *Puppy Talk*

WONDERDOG

The Science of Dogs and Their Unique Friendship with Humans

JULES HOWARD

PEGASUS BOOKS
NEW YORK LONDON

WONDERDOG

Pegasus Books, Ltd.
148 West 37th Street, 13th Floor
New York, NY 10018

Copyright © 2022 by Jules Howard

First Pegasus Books cloth edition November 2022

All rights reserved. No part of this book may be reproduced in whole or in part without written permission from the publisher, except by reviewers who may quote brief excerpts in connection with a review in a newspaper, magazine, or electronic publication; nor may any part of this book be reproduced, stored in a retrieval system, or transmitted in any form or by any means electronic, mechanical, photocopying, recording, or other, without written permission from the publisher.

ISBN: 978-1-63936-262-2

10 9 8 7 6 5 4 3 2 1

Printed in the United States of America
Distributed by Simon & Schuster
www.pegasusbooks.com

For Biff

For Chan

For Oz

For every dog known

Contents

Prologue 9
Introduction 13

SECTION I: SIT, STAY 23

Chapter 1: From streets they came 25

Chapter 2: Emancipation Day 45

Chapter 3: Sacrificed for science 59

Chapter 4: The Brown Dog Affair 73

SECTION II: FETCH, RETRIEVE 91

Chapter 5: Alpha, beta, doubter 93

Chapter 6: Skinner, unboxed 113

Chapter 7: The cognition ignition 123

Chapter 8: How nature met nurture 137

SECTION III: MEET, PLAY, LOVE 155

Chapter 9: What is it like to be a dog? 157

Chapter 10: Flip, the switch 175

Chapter 11: The power of play 193

Chapter 12: To see love coming 215

Epilogue: … and see love depart 239
Acknowledgements 245
Research notes and further reading 247
Index 279

'What cannot be denied or evaded is that this science has a moral dimension. How we study animals and what we assert about their minds and behavior greatly affects how they are treated, as well as our own version of ourselves.'

– Dale Jamieson

Prologue

Before we begin, a quiet reminder that, for the vast majority of human existence, there was no such thing as a home as we know it today. That, for only the past 400 or so generations, our ancestors have known what it is to construct a base – to use mud, stones, wood and bricks to make something that erodes into soil or sand more slowly than its surroundings. Dogs have been a big part of our lives during this period, but it is only very recently that so many have been invited into our homes to become part of the family. House-trained, so to speak. To convey how short a period this actually is in the grand scheme of things, I turn to the time-honoured geological tool of communication – the toilet-roll timeline – to help explain.

Let us apply the toilet-roll metaphor to the human story, by imagining that the first sheet of a freshly unwrapped toilet roll has upon it some of the earliest representatives of the hominid lineage five million years or so ago. In this context, most of the toilet roll involves activities that are more ape-like than human. Toilet sheet after sheet, the stories of those early hominids are written on the paper; on those fragile squares our ancestors chase, hide, migrate, grunt, laugh, frolic, politick in their social groups, much like we see chimpanzees do today. That's life, to a close approximation, for ape-kind. In fact, it is only about halfway through the toilet roll (say, 200 sheets in) that members of our lineage begin to show an affection for anything else. It's at this point that our ancestors develop an affinity for stone tools, something of an artistic passion they clearly come to enjoy. If the second half of the toilet roll reads like a song, it is mostly – to all intents and purposes – one long, single-note, haunting homage to the plasticity of stones. For most of our existence as a species, that has been our de facto behaviour: fooling around with stones and primitive spears. That's what we did. That's who we were. Sheets 250 to 300: stone tools. Sheets 301 to 350: stone tools. Sheets 351 to 400: stone tools. Sheets 401 to 450, the same. But then, just as we get to our final single sheet of loo roll: change. Just as the grey cardboard becomes visible under that final sheet, there on that single page in time: progress.

On that final sheet of toilet roll, many human civilisations across the world began a sudden wave of invention – of agriculture, architecture, governments, writing, civilisations, sewers, schools. That sheet is the so-called Neolithic Period. The Neolithic Period is, there or thereabouts, when dogs joined the party in a big way. Though they had been around our encampments for perhaps thousands of years, this was when humans began to take notice of them, pulling them closer and closer into human cultures and, in turn, being pulled closer and closer into theirs.

From the bottom of that last square of toilet paper, run your fingers gently upwards towards the grey cardboard cylinder. Feel the lives and livelihoods of 10,000 years of ancestry. Picture the experiences and daily lives of dogs, most making a living from scraps, throwaways, leftovers. Some kept as pets. Some trained to fight. Some bred to hunt, to corral, to retrieve. Closer, closer, closer still, your fingers move towards the end of the sheet, nearer and nearer to the cardboard, where the story of time meets now. When your finger reaches that final centimetre, stop. Look closer, and measure out 2mm, almost a hair's breadth. This tiny fraction of the entire toilet roll is a preposterously small amount of time in the grand scheme of things, but that's when the dogs moved in. That's when, in many parts of the world, a great many millions of dogs were invited into our homes to live alongside us. Actually invited. When they were fed and watered. When they were taken for walks. The period in which their being a part of the family became standardised. The bit when they sat on request. When they sat on our laps as we watched TV. When they slept in our beds while we drank coffee on Saturdays. When they were trained. When they were provided with hospitals. When they became insurable items. When they became housemates, friends, companions.

Everything that defines our modern relationship with dogs as you or I know it happened in this final millimetre of human existence. Everything you and I know about the minds of dogs was discovered during this infinitesimal fleck of time.

Imagine the secrets we have yet to discover about one another in the years to come.

Introduction

Life is precious, they say. Life on Earth, more so. That we live on a lively spinning sphere, revolving around a burning star with a name only we know, is staggering. That animals have evolved, through a natural process without any Upstairs Planning in the least, is too unlikely a concept for many to grasp. I sympathise, sometimes. Life truly is too beautiful. Incredible, really. And the staggering thing about life on Earth is how life begets life. How animals make it their duty to bumble into one another. That their ways of life frequently combine with others'. That there is predation. Competition. Nepotism. War. And peace. That the fortunes of one species can lead to the waxing and waning of another. That there is mutualism, where an organism works with another and both parties better their life-chances as a result.

Such as the coral polyp that provides a safe space for photosynthetic algae to divide within for a rental fee paid in energetic return, or such as the pollinating midge and the tiny flowers of the cocoa plant. That there is commensalism in nature (where one species happily takes advantage of another at no cost to the other party) and parasitism (where the cost can be large). For me, as someone who has written about animals for more than twenty years, the delights of life are in the interactions between individuals and their species. That's where the stories are.

And so, over the years, I have celebrated panda sex, charted the fossilised unions of extinct animals, watched in awe at stickleback trysts. I have counted the sexual games of toads and frogs; logged their bouts, the winners, the losers. I have been mother to hundreds of baby spiders, threatened with extinction – releasing them into the wild as if they were eight-legged children leaving home for university. I have gazed upon microscope slides of mites that live on slugs, and of slugs that live on slugs. I have seen bonobos having sex with, well, most things, prize-winning horses having sex for money, and tiny rotifers not having sex for 50 million years or more. And, well … you get the idea.

Throughout this period, there has been one relationship between species that blares out like a siren into the natural world – a siren I have found hard to ignore. It is a relationship unlike any other. I am referring, of course, to humanity's relationship with dogs.

Almost everywhere there are humans on planet Earth, there are dogs – strange canine interlopers who found their way into our lives thousands of years ago and have yet to leave. They are the first animal that humans domesticated, beginning perhaps 30,000 years ago, yet they are so staggeringly different from other domesticated animals. For starters, in their once-natural state, dogs are dangerous predators. Far from being relatively easy to confine – like chickens – dogs are wily, stealthy and athletic. Crucially, dogs helped us connect with the wild in a way that our human senses do not allow. It was dogs' noses that first sensed dinner; when we hunted, it was their trail we followed. (Never, in the history of the universe, has a sheep led a team of spear-clad hunters to a meal.) And then there is the connection we feel with them. If you have picked up this book, you are likely to know this extraordinary connection too. Dogs are our friends in a way that most other domesticated animals are not. They have captured our hearts and minds for millennia. Theirs is a strange and unique magic. Together, we make sparks. This is not parasitism. It is not commensalism. It is not classically mutualistic, either. It's something else.

Strangely, this unusual relationship has not always been of much interest to zoologists. For decades in the twentieth century, dogs were considered unworthy of rigorous study. Academics deemed them broken by humanity's influence. They argued that the very act of our cross-species union muddied their evolutionary back story. Far better to seek out the wild account spawned by nature – the grey wolf, red in tooth and claw – than the 'dumb wolf' that hoovers scraps from under our kitchen tables, they contended. This snobbishness about dogs became widespread – I certainly remember this being the attitude when my zoological studies began in the 1990s. To the old guard, dogs were frowned upon as animals worthy of scientific attention. Focusing on dogs to understand the evolved behaviours of wild canids (the mammal group that includes fox, domestic dogs, coyotes and wolves) was like trying to understand the adaptations of a chicken's egg by studying the crumbs of a wet cake. Too late, they claimed. The ingredients were forged too long ago. Humanity had corrupted dogs, we were told. We had bred the wild out of them. Enjoy them, sure, but there was no point in studying them. In time, this attitude would change, morphing into something else entirely. It would change what we know about animals.

In recent years, many biologists have returned to dogs. In dogs, they argue, we can see elements of behaviours or characteristics that natural selection has whittled into shape through thousands of years of living wild. Crucially, though, in dogs we can see new behaviours, new cognitive skills, new ways of thinking imposed upon them by our close association. In Victorian times, many scientists studied animals to understand the mind of the Creator. Today, we see in studies of modern dogs evidence that that Creator is us. A creator (note: lower-case) who acted, for the large part, unthinkingly, but also a creator who did not work alone. In fact, for most of their history, we now realise dogs really did choose us as much as we chose them. Dogs have the history of our union built into their genes. But somewhere or other, in fleeting glances, we see this union in ourselves too. In our history. In our sociality. Perhaps, in our genes.

The last two centuries have seen an enormous change in the strange relationship between human and dog. But another turbulent time is beginning as you read these words. According to Statista, the consumer data specialists, right now dog populations are on the rise across many of the world's Western nations. Since 2000, the USA's dog population has risen by 20 per cent: it now stands at 89.7 million dogs and counting. In the UK, the trend is also pronounced: according to annual surveys by the PDSA, there has been a 20 per cent rise in a single decade, with the figure now standing at 9.9 million. Germany has a similar figure to the UK, with 9.5 million dogs, and tops the chart of dog-loving EU countries. Overall, the population of dogs in the EU stands at roughly 65 million. That figure is also growing: one survey suggests that the number of dogs across Europe is growing at a rate of 3 million each year. Populations of pet dogs are also on the rise in Australia: in 2016, there was approximately one dog for every five people across the country – 4.8 million dogs in total – but this figure is rising by about 200,000 each year. The trend is perhaps most marked in Canada, where a 20 per cent rise was observed between 2014 and 2016, with Canada now home to more than 7.6 million dogs. Statistics like these show that dogs are becoming an increasingly important part of people's lives.

Partly because working from home allows more families to keep a dog responsibly, the Covid-19 pandemic saw dog numbers continue upwards. According to Google Trends, comparing April 2019 (pre-pandemic) to April 2020 (when many countries were experiencing their first lockdown) searches for 'puppies for sale' approximately doubled. The country spread of searches was clear: in the USA, in Canada, in the UK, in South Africa, in Australia, Ireland, New Zealand. This surge in interest translated quickly to the price of puppies: in the UK, research undertaken by The Dogs Trust suggested that the price for some breeds had doubled or almost tripled during this time. A dachshund puppy before the Covid-19 pandemic was, on average, £973. After the first lockdown, it was more than £1,800. After the second, it was nearer £3,000. After the third, it was nearer £3,500. Clearly, this sharp price rise caused concern for many. Puppy farms – where puppies are mass-produced for profit, often in the most dispassionate, cruel and unhygienic of ways – attempted to fill this void illegally.

To ensure the best relationships during this period of dog-population growth, we need the best information going – the best insights, the best impartial findings. We need to help the scientific research (often hidden behind incredibly expensive paywalls) find a mass market. We need accessible science, in other words, which is one reason I began writing this book.

But there is another reason. I am aware that there are many books about dogs, their behaviours and their impressive cognitive skills. In fact, many of the authors of these books have been a great inspiration to me over the years. These books often focus on *what* the dog is thinking, on *what* the dog knows and *what* the dog does not know. Many are accessories to training regimes – guides for what to do and what not to do with your dog. They are superb, well-researched, technical guides to 'knowing' a dog. But my aim with this book is different. My feeling is that, in order to gauge successfully where the human relationship with dogs may go from here, we need to see where we've come from. We need to remind ourselves *how* we came to know the mind of dogs. Only then can we prepare and plan for where we might go next.

I would argue, with a nod to my own pomposity, that understanding animals is a bit like understanding the solar system. A book about the moon is interesting, sure. Vital, even. But the story of how we got to the moon adds a different context – that is a story of achievement, as emotional as it is technological. Both stories have value, but only told alongside one another can stories like these spur us on to even greater achievements, to be a better species. In this context, history really matters.

I would argue that it's the same with dogs. Knowing what dogs *do* and perhaps what they *know* is one thing, but knowing how we have come to comprehend such things about their minds is another thing entirely. It puts into context our understanding of them, and it forces us to acknowledge that what we know about dogs might change in future, as more facts and insights become available. In fact, our relationship with dogs is almost certain to change again, hopefully in a way that is beneficial to both species.

The scientists (alongside the dogs) are particularly important characters in this book. Knowing them helps us to understand the junctions, the circuits and the parameters of intellectual travel. These individuals help us to understand that much of what we know about dogs is framed within the mind of the human experimenter, a species that is changing at its own pace – that is changing its own perceptions of place – in the modern world. My belief is that knowing all these things will help us be better companions to dogs, and help us succeed in making the lives of dogs as happy and as healthy as they can be.

The message of this book is straightforward. It is simply that the more compassionate we have become in our explorations into the minds of dogs, the more intelligent they have shown us to be. It's that simple. I have come to see that dogs are a message to all of us in how to study nature, in how to throw open the gates of evolutionary thought, in how to gauge our place in the world, in how to make this planet a better place, perhaps, for all species. It is a story of how the quality of science improves when we treat animals with empathy. And how the greatest feats that dogs have shown themselves capable of have been at the hands of humans who know and love them. Perhaps I'm biased, but there is a certain beauty to this observation.

Now, talking of biases, I am honest about the biases in this book. Clearly, one large bias I am carrying is that I am hopelessly in love with dogs – an affliction that makes writing about them as cold research subjects challenging at times. I am aware this limits how independent and balanced my outlook on this species can be. However, like the behavioural scientists Alexandra Horowitz and Marc Bekoff (this book owes a debt to both), I am happy to argue that it is possible to – occasionally – dip into anthropomorphism yet keep oneself well within the boundaries of good science. To quell this bias, I have pulled upon the intellect of minds far less woolly than mine. In fact, many veterinary professionals, psychologists, ethologists, neurologists, historians and others have allowed me to step into their areas of research so that I can best convey to you, the reader, the fruits of their exploration. I hope not to let them down.

I mention biases mainly because, try as we may, separating science from human biases, human cultures, human moralities, ethics and idylls is extremely challenging. This applies, perhaps more than anything else, to the science of dogs and what goes on in their minds. Thus, what you are about to read is as much a story about humans as it is about dogs. About how we treated dogs like objects at first, then as inmates, then patients and, finally, learning-companions, partners and something akin to metaphorical co-pilots in a rocket flying past the moon and on to cosmic pastures new.

Not all of the stories in this book are pretty, however. Particularly in its infancy and indeed into the 1960s, dogs were often treated in the most miserable and disturbing ways by research scientists. Sensitive readers should note that I have kept much of the gory details out of the main body of the text, flagging them up in footnotes and in the Notes and Further Reading section at the end of the book. Though there was a temptation to remove this information entirely from the book, my hope is that some readers can view the suffering of dogs from a modern-day vantage point, seeing how far our relationship has developed and reminding ourselves where we have come from and should never return.

The book begins with Darwin. We first explore the Victorian era and what exactly dogs represented to science and society. We look at the earliest experiments: at the dogs carrying signs or trying to manipulate big sticks through small fences and failing every time. We consider how science came to know their sense of smell. Of touch. Of memory. Of taste. We explore the mistreatment of dogs in science during this period, and the rise of animal rights organisations, many rebelling about the atrocities forced upon dogs in the secretive laboratories of medical institutions. We chart rabies. The decline of street dogs across much of the Western world. We move from Darwin to Dickens. To dog shows, pedigree breeds, dogs turning spits. From here, we journey through

Thorndike, Pavlov and Skinner, scientists who thought every quirk and facet of dog behaviour could be trimmed down to simple conditioned responses – something akin to the notion that, if it feels good, then do it again.

From here, in the mid-1900s, we see the developments of three competing fields of science that often used or depended upon dogs for their springboard moments: psychology, behavioural genetics, neurobiology. These diverse fields of science saw in dogs a suitable, worthy, study animal through which we could learn about animal intellect, emotions, feelings and, of course, the science of cognition – how, exactly, animals acquire understanding through senses, thoughts and experiences. The input of dogs on these fields is largely forgotten, so it feels right to bring them back to the fore.

In the final third of the book, we discover the fruits of more modern-day behavioural research: that dogs recognise themselves as individuals, through their play behaviours, through their responses to us, through our everyday interactions with one another. Then, in the final chapters, we map the first decades of this century, a period during which our knowledge and understanding of dog cognition has increased perhaps tenfold. These exciting decades have seen a flourishing of the field of anthrozoology, of citizen (dog) science, of experiments where the dogs are not subjects but playmates and canine collaborators – true wonderdogs.

This book is called *Wonderdog* for a reason. Many of the most staggering recent discoveries you are about to read came at the hands (or paws) of family dogs, of dogs with names. Dogs like Oreo, who defied one of the greatest minds in psychology by understanding the importance of a pointing human finger. Dogs like Flip, the stray taken home by a research scientist who went on to inspire a global wave of dog cognition studies. Dogs like Marla, a gorgeous sheepdog addicted to human company, courtesy of a handful of insertions upon the genes that code for sociability. And Rico and Chaser, the laser-sharp, bounteous collies who could recall the names of hundreds of toys, and all in the name of fun. Each of these characters is nothing short of a true wonderdog; each helped lay a path towards discovery that scientists followed; each changed the way we viewed the world – how we saw our place in nature and our connection with other life forms on Earth.

Life is precious and so are relationships. Dogs have shown us so much. It is my belief that there is much more they can bring out in us; we just have to keep asking the right questions in the right way. I hope this book inspires you, the reader, to do that.

SECTION I
SIT, STAY

CHAPTER I

From streets they came

> 'The known is finite, the unknown infinite; intellectually we stand on an islet in the midst of an illimitable ocean of inexplicability. Our business in every generation is to reclaim a little more land.'
>
> – T. H. Huxley (1887)

Our story will have many players. A cast of Americans, Russians, French, Scandinavians, Hungarians – all with science on their minds and in their notebooks. Throughout the pages of this book, their discoveries, theories and ideas will move like waves between continents, leaving eddies in their wake in which culture, belief and passionate opinions swirl, entangling and coalescing or bouncing apart violently.

But stories like this one must begin in a time and a place, and ours is mid-nineteenth-century London. An age in which a fashion for dogs was blossoming in the middle classes, where Charles Darwin's ideas had recently forever changed our understanding of humans and animals, where societal injustice was being challenged in new ways – a time at which the life sciences were becoming ever more a laboratory pursuit.

To know this place, both scientifically and culturally, is to get a feel for how far the story of dogs and science has progressed, to see how far we have come.

Our scene-setting exercise begins with Charles Darwin.

Take a moment, if you can, to glance over at a nearby dog. Think about what you share in common. Take in the up–down jaws of their skull. The paired nostrils and ears. The in–out mechanism of the breathing. Look at the muscular tongue. Notice their eyes looking back at you. Look at the intensity, shining back. The interest. Gaze into their pupils. Take in the eyelashes. If you're lucky, you might share a smile.

Now, if you can, lean in for a stroke. Feel the bones in the legs first. Notice the arrangement and how they mirror those in your own limb bones. Start with the heavy bones at the top of the legs: the humerus (forelegs) and femur (hindlegs). Work your way down to the paired bones that connect to them: the ulna and radius in the forelegs and the fibula and tibia in the hindlegs. As with your own, these bones have within them both yellow and red bone marrow. They are factories that produce blood cells and maintain the body. They are what keeps your dog alive.

Move onwards. Run your fingers down your dog's neck, feeling for the seven neck vertebrae that nearly all mammals possess. Then move your hands lower and guide them in the channel between the shoulder blades (scapulae – again, you have them) and down the spine. Put your hand in front of its mouth now. Feel its muscular tongue give you a loving lick. Observe the arrangement of the teeth – the incisors, the canines, the molars. Like yours, these are likely to be adult teeth; its milk teeth were lost long ago, probably swallowed while eating. Finally, go paw to paw. Feel with your digits, the digits of your dog. The same familiar arrangement, with dewclaw as thumb.

It is striking to imagine that before the 1800s scientists lacked good evidence to explain why our bodies are so alike – why the bones of humans and dogs and, well, most mammals are so similar in their arrangement. The traditional argument, based on Aristotle's ideas two millennia ago, was that animals were arranged in a kind of ladder of progress, with lower animals (starfish, sea-squirts and the like) near the bottom, and animals that had achieved perfection (namely, us) at the top (beneath God and then His angels, of course). In this primitive organogram, dogs were often two or three tiers down from humans, somewhere alongside elephants, camels, horses and dragons.

This Bible-led interpretation considered the first dogs as a resource made by God to serve 'mankind'. It saw those earliest dogs as a kind of seventh-day sheepdog that lived to serve but then became corrupted over time into different breeds ('races') by jobs locally required of them. During the late-eighteenth century, the 'recognised' European breeds included Siberian, Icelandic and Lapland herding dogs and Pomeranians; greyhounds and mastiffs for sight; hounds, terriers and spaniels for their noses. Due to their numerous breeds, dogs became a useful study species to naturalists of the time, who were eager to understand the diversity of life and what might account for it.

'Of all animals, the Dog is also most susceptible of impressions, most easily modified by moral causes, and most subject to alterations occasioned by physical influence,' wrote the influential French naturalist Georges-Louis Leclerc,

Comte de Buffon. 'His temperament, faculties, and habits, vary prodigiously; and even the figure of his body is by no means constant.'

Working in the late 1700s, Buffon played a vital role in helping comparative anatomy reach the mainstream. By lining up animals and comparing their organs, their bones, and other body structures, Buffon hit upon a scientific way of studying the order of nature and contemplating ideas such as adaptation over time to local environments. 'He was not an evolutionary biologist, yet he was the father of evolutionism,' wrote the twentieth-century biologist Ernst Mayr, paying homage. 'He was the first person to discuss a large number of evolutionary problems, problems that before Buffon had not been raised by anybody ... he brought them to the attention of the scientific world.'

Buffon's work touched upon inheritance, taxonomy; it suggested that the Earth had a deep and rich history – a controversial notion at the time; he even noticed the 'struggle for existence' in nature, a key driving force behind how organisms change over time. But Buffon couldn't get past the idea that dogs and other animals might not have been placed here by an all-knowing God at the beginning of time. It was too radical an idea. It was on that issue, the so-called 'immutability of species', that Charles Darwin would eventually take the plunge.*

* I mention only very briefly Alfred Russel Wallace in this section of the book. This isn't to belittle his contribution to the discovery of natural selection – after all, without Wallace's insight, perhaps Darwin might never have had the courage to publish – but rather it's a reflection of the power of Darwin's well-constructed and coherent argument, outlined in *On the Origin of Species*. It was this, not the discovery per se, that captivated the public at the time.

Darwin outlined in 1859's *On the Origin of Species* a natural mechanism through which species could change over time – this was natural selection, where individual variations within a population are acted upon by the selecting agents of ill fortune, and where successful strains prosper at the expense of failed ones, and the world becomes populated with generations of successful mistakes – or 'the plodding accumulation of error', as science writer Steve Jones puts it. Darwin concluded that the shared features of animals came about because of shared ancestry; that the features we share with other animals often resemble one another because we have inherited them from a single collective grandparent that lived millions (or sometimes thousands) of years ago. With a hint of his trademark flourish, he put it like this: 'As buds give rise by growth to fresh buds, and these, if vigorous, branch out and overtop on all sides many a feebler branch, so by generation I believe it has been with the great Tree of Life, which fills with its dead and broken branches the crust of the earth, and covers the surface with its ever branching and beautiful ramifications.' This is, of course, why we have the same skeleton as a dog: because we both inherited frames from a shared common ancestor, a diminutive little mammal that lived long ago in the shadow of the dinosaurs. A mammal that – if you travelled back in time and did not know of its significance – you would almost certainly forget you had ever met.

Let us consider this Jurassic character for a moment. Fossils suggest it was probably an insectivorous furball with a basic mammal skeleton, like the one you and your dog now live within, in modified form. This shared prehistoric grandparent of ours almost certainly had nipples, three tiny ear bones, and an umbilical scar; it had a beating heart, liver, kidneys, paired lungs and pretty much every mammalian gland you have pumping and squirting in and upon your body right now. All mammals today share these features because we inherited them from the same flighty, highly sensitive, Mesozoic toe-rag.* It is from these animals that both dogs and humans are derived. This is written in our bones. Later, long after Darwin, we would discover it written into our DNA too.

The popular perception is that Darwin and his ideas caused uproar to polite society at the time, but what is surprising is how little disturbance his ideas actually caused most people. 'Within Britain, *Origin* did not so much initiate a crisis as conclude a major piece of unfinished business from the 1830s,' writes James A. Secord, Director of the Darwin Correspondence Project. 'With significant exceptions, as Darwin acknowledged, reviewers treated his arguments patiently and in good faith.' The positive reviews of the time talk of 'markings of change' (Scottish author Robert Chambers), of paths of enquiry 'full of promise' (English philosopher John Stuart Mill), and of a book that is a 'rational revelation of progress' (French scholar Clémence Royer).

Not all the reception for the book was quite so rapt, of course. Famously, one of Darwin's closest allies Thomas Henry Huxley (known affectionately as 'Darwin's Bulldog') went toe to toe over the book with one of his greatest opponents, the Bishop Samuel Wilberforce. After one particularly well-attended lecture at the Oxford University Museum, baited by his friends, the Bishop socked it to Huxley with a question about whether he'd rather have a monkey for his grandmother or a monkey for his grandfather, to which Huxley responded that he'd much rather have a monkey for a grandfather than a man with such a lacklustre sense of humour as he, or words to that effect. The true exchange is lost to

---

*Mammals are also known for their whiskers. In humans, our whiskers are long gone, but histological studies performed on cadavers suggest that 35 per cent of people probably still possess the facial muscles to make them twitch.

history. 'Let them rage!' said German zoologist Carl Vogt of vicars and priests.*

These minor scuffles aside, the ease with which Darwin's ideas oozed into society tells you a lot about the style of the book itself. It says a great deal that the book can be picked up today and remain, in parts, wholly engaging. This is because Darwin wrote with his audience in mind. To add extra weight to his ideas, Darwin chose not to focus on faraway animals like baboons or fruit-bats or exotic badgers or tigers. Instead, he expressed his ideas by pulling on the everyday animals that people understood best. Thus, by opting for the unremarkable, Darwin's ideas became all the easier to apply. He regularly drew upon dogs (though not nearly as much as pigeons – Darwin was a renowned pigeon-fancier) and notices, for instance, that pointing, circling around a flock and retrieving are all behaviours that wolves achieve in their wild states and speculates many times on their ancestry. To Darwin, the ancestry of dogs remained something of a mystery. He could see the power that variation brought to his theory – that nothing evolves without small (or sometimes large) differences between members of a species or population (a longer femur or shorter tibia here, a sharper nose there, *etc.*) – but the problem was that dogs were just so magnificently varied in their shapes and forms in a way that other species were not. There were mutant bulldogs, with bulbous heads and misshapen, undershot jaws, turnspit dogs (whose job it was to run on

---

* Darwin later wrote, in *The Descent of Man*: 'For my own part I would as soon be descended from that heroic little monkey, who braved his dreaded enemy in order to save the life of his keeper; or from that old baboon, who, descending from the mountains, carried away in triumph his young comrade from a crowd of astonished dogs – as from a savage who delights to torture his enemies, offers up bloody sacrifices, practises infanticide without remorse, treats his wives like slaves, knows no decency, and is haunted by the grossest superstitions.'

wheels that turned spits) and greyhounds, which he saw as 'the perfect image of grace, symmetry, and vigour'. Darwin hypothesised that this much variation could not possibly come about through the domestication of one single species. Instead, he hypothesised that dogs had their origins in numerous canid species (particularly jackal and grey wolf) that had crossbred throughout history. Of course, as we shall later explore, we now realise this assertion is false.

Dogs are mentioned fifty times in *On the Origin of Species*. This is, in part, because Darwin kept dogs and always had a professional and private interest in them – a love of them, really. But, by talking so frequently about the history of dogs, Darwin surely knew he could make a connection with the reader. He is likely to have seen the recent changes in society, with dogs in the mid-nineteenth century increasingly being brought into homes and kept as pets, courtesy of a newly burgeoning middle class. It's arguable that Darwin knew dogs would be a good way to gently introduce big ideas to polite society. And, in this aim, he succeeded. In the words of historian Emma Townsend, author of *Darwin's Dogs*, he brought 'evolutionary theory right to the hearth rug of the Victorian home'. And, with the family dog at rest, what a home it was.

In the UK and across Europe, the human–dog relationship was changing at great speed. A hundred years previously, in the eighteenth century, the idea of dogs being given legal protection from cruelty would have been largely inconceivable. In fact, the idea of keeping dogs in the house, all snug and warm, was almost unimaginable to all but highbrow ladies at the upper echelons of high society. But this was Victorian Britain, and a new cultural norm was being established by a blossoming middle class doing rather well out of the Industrial

Age. Dogs were a fancy for this swelling demographic. As family pets, they came into homes and were cared for, catered for, loved and adored. Hand in hand with this shift came an intolerance towards animal abuse or torture. According to the anatomist and historian Alan W. H. Bates, 'this was mostly due to London's changing demographic: in the crowded capital, the well to do could not avoid witnessing the brutal treatment of draught animals and livestock.'

Literature played no small part in the speed with which this cultural pivot towards dog ownership occurred. Magazines and periodicals of the time became obsessed with stories of dogs that were imbued with impressive human-like powers. Stories of genius dogs flourished. Magazines covered again and again the stories of dogs such as Greyfriars Bobby, the Skye terrier who apparently spent fourteen years guarding the Edinburgh grave of his human companion. The case of a Pennsylvanian bulldog, who observed his 'master' receive a medical splint for a broken arm and later brought in an injured street dog expecting similar treatment. Dogs who could count money and settle debts and wagers. Dogs who slavishly defended babies from wolves or petticoats from robbers. Among all this, there was Charles Dickens, a most Victorian lover of dogs. Perhaps the most celebrated dog-lover of all.

For a man said to have greatly swelled the national affection for dogs, Charles Dickens' personal accounts of dog-keeping occasionally read more like something from *The Beano* than *Bleak House.* Among his most memorable was Bumble, a Newfoundland whose obsession with returning home ahead of his apparent master was, for reasons unclear, infuriating to Dickens. Then there was Sultan, his Irish Bloodhound, who was excitedly led out in the garden as if receiving some lovely treat, only to face an armed firing squad ordered to kill him for biting a child the previous day. And, of course, there was Timber (or Timber Doodle), a white spaniel given to him in

1843 whose loose bowels and complete disinterest in sexual intercourse with other dogs brought upon him great disgrace. To add insult to injury (and to the apparent horror of Dickens), Timber eventually began expressing sexual feelings for a pet white rabbit. He really did become something of a trial for Dickens.

'Timber has had every hair upon his body cut off because of the fleas, and he looks like the ghost of a drowned dog come out of a pond after a week or so. It is very awful to see him slide into a room,' Dickens wrote in a letter to a friend, dated 1844. 'He knows the change upon him, and is always turning round and round to look for himself. I think he'll die of grief.'*

Mishaps like these aside, Dickens and the apparent ease with which he could inject dogs with character or dress them up in metaphor was compelling and in some ways unseen in literature up until this time. This novel fondness for dogs was picked up on by many. The author Percy Fitzgerald (1834–1925) wrote of his acquaintance: '[Dickens] takes the newly-enfranchised animal within the charmed circle of his characters, sets him down at the fireside and chimney-corner, and furnishes him with quaint reflections of the whims and humours of humanity, playing on them with delicate touches which seem almost earnest, until they really mount to the dignity of a character.'

Among Dickens' more famous dogs are *David Copperfield*'s Jip, Boxer from *Cricket on the Hearth*, and Merrylegs, the abused circus dog in *Hard Times* who, some argue, serves as a metaphor for the mistreatment of the working classes. And, of course, the ghoulish double act of *Oliver Twist*'s Bill Sikes and Bull's-eye, both complicated monsters of a dingy

* This was one of a number of Victorian treatments for dog fleas. Others included the application of kerosene, carbolic acid and whale oil.

underworld that would influence gritty crime-dramas for centuries to come.*

To walk through Dickens' idea of London was to find oneself surrounded by the sights and sounds of animals – of horses, cats, caged goldfinches, sheep, pigs and armies of hungry street dogs. In this cultural milieu, street dogs slept, they fought, they pulled apples from carts, they barked at horses, bared their teeth and licked their wounds. Yet, among these dogs there was a new character. These were different dogs. They looked different, for starters. They walked differently too. They were the dogs that were pampered and cared for by people who treated them as equals, as friends, as family. It was these people – a new, burgeoning middle class with time on their hands – that Dickens' stories resonated with. It was the same audience that had been so energised by Darwin's great revelations about shared animal ancestry, physiology and shared emotions with other animals. Like two great waves syncing up – one science, one story – the cultural ripples across society were impossible to ignore. Dogs really were on the up.

By the time Darwin's published works were gaining traction, Britain was in the midst of a kind of national hysteria for keeping dogs as pets. A sympathetic movement for dogs was growing, transmitted from park to park, public garden to public garden, club to club. Like a trend gone viral, dogs and their breeds very quickly became big

* In popular culture, Bull's-eye is often portrayed as a bulldog. In fact, Dickens makes no reference to its breed in *Oliver Twist* – he is apparently 'a white shaggy dog, with his face scratched and torn in twenty places'. I am informed by Alison Skipper, vet and dog historian, that this idea probably began with the original illustrations by George Cruikshank, which depict Bull's-eye as the old, athletic type of bulldog: 'I think this is because the cultural trope of the time of the bulldog as a seedy creature of the urban underworld was just too good to miss,' she explains.

business – status symbols, must-have accessories, canine keepsakes and sweethearts.

The street dogs (of which there were many) watched on, perhaps with envy. These dogs had a different kind of viral madness to contend with. It was a disease that would see them removed from the streets for ever, leading future generations to completely forget they had existed. Though our focus in this chapter is on dogs of the home, these street dogs deserve a quick digression.

What exactly are street dogs? What does this name conjure up in the modern age, in cultures all around the world? How do street dogs best reflect our modern idea of what a dog is? Questions like these are worthy of consideration in this early stage of the book, not least because they force us to see them as ecological agents in their own right, with history stretching far further back than the nineteenth century. But these questions also have value because, in the modern age, dogs like these largely represent the common condition for dogs on our planet. Theirs is a niche that, in many parts of the world, continues to boom. In all, there are estimated to be 900 million dogs on this planet and a staggering 83 per cent of them are street dogs, often known simply as 'strays' – a catch-all term used to describe dogs that lack a formal owner.*

---

*The word 'owner' brings with it many connotations and, understandably, many readers will prefer I don't use it. However, I have chosen to use it in certain parts of the book, simply because it describes, legally, the relationship that we share. Though I abhor the idea of dogs being property, that is clearly how they are viewed in a court of law. In future editions of this book, I hope that dogs will be afforded rights that represent more closely the relationship we share; at that point, I'll gladly rephrase.

Your dog, if you have one, is likely to fit into a much smaller category. It is, to use a technical term, an 'owned-restricted' dog.* Owned-restricted dogs are lucky dogs. They are dogs that are fully dependent on the humans with whom they cohabit. All of the essential needs of owned-restricted dogs are provided courtesy of human hands – the food placed in a bowl, walkies, cuddles, bath-times, trips to the vets, the lot. That's not to say that all 'pet' dogs fit under the 'owned-restricted' banner, however. There are some dogs that people would still consider part of the family but that roam the neighbourhood, wandering in and out of places freely, sometimes garnering local adoration from friends or fear from strangers, depending on their temperament. These are termed 'owned-unrestricted' dogs.

Stray (or 'village') dogs make up the bulk of dog populations that fill nearly all of the world's continents, but the category itself is something of a broad church. Strays include dogs that may once have been born into a human household from which they have been ejected, or that have been born on the street but have since maintained regular interactions with humans either through scavenging or through handouts. In many cities around the world, these are the dogs that roam the streets, sometimes in loose aggregations. These strays can be very tolerant of people but many may be highly fearful or violent and aggressive around people.

The final category of dogs are much harder to study, to quantify or even to see. They are the feral dogs that live in the wild with no help from humans by way of food provision or shelter. These dogs have, in an evolutionary sense, lost their connection to humans. The lack of a human presence in puppyhood, particularly, sees these dogs lose their attachment

---

* These categories are those used by Boitani *et al.* that take into account the level of human–dog interactions. These are broadly comparable to those used by the World Health Organization (1988).

for us and often go on to see in humans little more than a source of great fear. Feral dogs include the dingoes of Australia, who parted ways with humans at some point in the last 6,000 years or so.

Of all these different categories of dog, by far the most numerous are strays. Across the world, these dogs live out their lives, achieving in opulent abundance what biologists like to call the Four Fs: fighting, fleeing, feeding and ... fornicating.

In recent years, village dogs* have proved particularly interesting to dog science because they provide an opportunity to consider how dogs might have behaved before the pampering began a few hundred years ago. They allow us to look back in time and, perhaps, to glimpse what our earliest relationships with one another may have been like. To see the ancestral environment that drew us together.

It is something of a fallacy that there was a great moment in history when human and wolf met on a mountain somewhere, a bronze collar placed around a neck, human hands licked for the first time. That the wild could somehow be tamed in a single mawkish moment, most scientists agree, is highly unlikely. Instead, the vast majority of dog scientists argue that wolves came for the scraps first, and then went from there. Their evolution began not at the hands of humanity, in other words, but in the handouts.

This behaviour – this link to human waste – is apparent in village dog populations the world over. Most scavenge from rubbish dumps, steal from bins or occasionally beg for handouts. Others go for less traditional foodstuffs. Most notably, there are the many millions of dogs who are partial to human faeces. So enticed by faeces are village dogs that

* I use the term 'village dog' from this point forth partly to follow the lead of the authors quoted in this section but also because 'stray' feels like something of a loaded term.

some scientists believe our toilet habits may have been a big factor in our enticing those earliest wolves.*

In the tropics and subtropics, village dogs are common – common enough that you are likely to see them in the backgrounds of your holiday snaps, engaging in all manner of F-related activities. Locally, these village dogs have their own names – the Canaan dogs of Israel, the Carolina Dogs of south-eastern USA or the pariah dogs of India, for instance. But some village dogs are very isolated locally and have roots that go deeper into the dog family tree. Among the most well known are the New Guinea singing dogs and the Kintamani dogs of Bali.

The pioneers of studies into village dogs and their relationships with humans around the world are undoubtedly wife-and-husband biologists and dog trainers, Lorna and Raymond Coppinger. The pair didn't set out to study village dogs, it was just that wherever they travelled to observe specific dog breeds in their natural habitats, or to attend conferences and the like, they noticed the strays at the hotels, the lodgings, the airports, the streets.† This, they eventually decided, would be their calling. In their generalised observations of village dogs, the pair concluded that village dogs aren't often more than 9kg, that they are relatively non-aggressive and that they show limited fear of people. Many, they argued, are able to make a good life for themselves

---

* In one study by scientists at the University of Addis Ababa (Ethiopia), 20 per cent of the diet of village dogs consisted of human faeces. In a separate study, scientists studying the faeces-eating habits of village dogs in Zimbabwe expressed no surprise that such a source of nutrition was so regularly consumed, being that human faeces is 'comparable to the upper range of energy content for mammal tissue, vegetables, and fruit.'

† 'They were so much more interesting than what we'd ever done before,' Ray Coppinger told *The New York Times* in 2016. 'Here were animals that had their own unique kind of social behaviours. So we started to study them.'

in this human-curated biological niche. One estimate has it that about seven free-living dogs can make a living off a hundred people's worth of garbage.

I confess a similar interest in dogs like these. My own journals, kept while travelling early in my career, seem mostly filled with daily encounters with village dogs rather than the animals I was supposed to be writing about. Blotch, a fast-talking everyman kind of village dog who, at 8 a.m. each day, would find time to sit like a gargoyle outside my door. Tick-tock, a poor poodle-like thing who had upon her ears an entanglement of engorged ticks that looked like jewellery. The Alliance, a gang of colourful misfits who would strut across the beaches in the morning, engaging in round after round of sexual intercourse, both homo and hetero. (Interestingly, in many months of observing the Alliance, there was never fighting among them, their impulses seemingly in throes to the mood of the day, the unseen fragrances of the females in tow.) The locals thought me mad, of course, a young zoologist taking an interest in these not-so animals. But these dogs were ever-present. And they quickly learned our routines. Each day, they would come to appear at the same times, making approaches in predictable ways.

In the Coppingers' travels, interviews with local (human) residents suggested that village dogs were treated with general (almost universal) aversion. Often, dogs were treated as vermin – a potential spreader of disease. Readily, they were framed as scavengers or, sometimes, thieves. 'In our interviews,' the Coppingers recount in *Genetics and the Behavior of Domestic Animals*, 'the cultural dislike for dogs was invariably presented first, followed by various individual modifications. These ranged from people who were disgusted by the thought of touching a dog, to others who thought dogs had some value as alarms or hunters of pests.'

One long-term study, undertaken by biologist Dr Sunil Kumar Pal and his colleagues in West Bengal, offers fascinating insights into the daily lives of pariah dogs. Here, according to their research, a given town might support hundreds of pariah dogs, each splintered into loose bands of family groups numbering five to ten. Often, dogs strolled alone. In a solitary way, they explored the backstreets, seeking new foraging opportunities – more like cats, almost. There were no tight-knit groups in pariah dogs. There is little evidence of obvious dominance hierarchies, including in males. When females come into heat, there is little bloodshed. Females may be courted by numerous males, many of whom she mates with. Often, post-copulation, one single male will choose to shadow her for a few weeks (forming what one might call a loose pair-bond) throughout her gestation. Occasionally this male may go so far as to communally raise offspring, regurgitating food for the puppies in the same manner that many other canids do. Here, among these human habitations, the social interactions of village dogs are loose and messy. This isn't an obvious law of the streets, except that polygamy rules. And crucially, there appears to be little or no human influence in the lives they lead. There are no people pulling the strings, attempting to modify or 'tame' the behaviours of these animals. No. Instead, the village dogs are more like an emergent property of human habitations. A new niche – perhaps less than 15,000 years old – in which dogs adapted quickly and heartily, getting in there before anything else.

These were the street dogs that once roamed London, New York, Paris, Rome and Sydney – many, if not all, of the world's major cities. That is, until rabies spread across the world, forcing us to act upon their kind with ruthless intensity.

As the final few decades of the nineteenth century began, and Darwin and Dickens' books and periodicals did their trade, the lives of street dogs in cities across the world were far from easy. The vast majority were frequent victims of cruelty by park keepers and the police. Stray dogs were often stoned by nervous members of the public, and those in the poorest state were quickly put out of their misery. The street dogs in the peak of health faced their own hardship: many were taken from the streets to take part in organised dog fights or slavishly employed to pull carts through city streets. Many churches employed 'dog-whippers', whose job it was to deter canine interlopers looking for their own version of salvation.* For many people in Britain at the time, just as the Coppingers found elsewhere on their travels, street dogs were little more than vermin. And this was before the arrival of rabies – a viral passenger passed in saliva, causing inflammation of the brain. The unpredictable behaviour of dogs suffering the disease – their frequent disorientation, incoordination and seizures, and their occasional biting of humans – was enough to send many cities and towns into a spin.

A swift rise in cases across cities in the Western world spurred governments, councils and civic authorities to take action on street dogs, sometimes in the most unscrupulous ways. Many cities in America had all sorts of problems. At around this time, for instance, the *New York Daily Times* lashed out against unmuzzled dogs that 'swarm in all the streets, obstruct the pavements, make night hideous with their howls …' The paper's rallying cry came six years into a city-wide campaign involving the clubbing and drowning of thousands of street dogs. During especially hot summers, when rabies was particularly prevalent, the New York police force also made use of a civilian workforce, offering a bounty of fifty cents for each unmuzzled dog taken to a

* The last recorded professional dog-whipper was a man named John Pickard, who was appointed to Exeter Cathedral in 1856. Today, the role continues in a purely ceremonial form, for processions and other significant occasions.

local police station for disposal. In this way, thousands of dogs met their fate at the hands of a hard-up public, presumably leading to enormous numbers of dogs suffering in the most awful ways.

In London, partly because of the dog's rising status as an animal worthy of rights, not all street dogs were dispatched in this way. Instead, many of the healthiest were taken to the newly built 'Dogs' Home' in Battersea – the brainchild of welfare activist Mary Tealby who, after losing a personal battle to save the life of a starving dog, resolved to never let it happen again. Tealby was something of an emergent cultural phenomenon: a powerful, committed, energetic campaigner who also happened to be a woman. She was a symbol of wider societal change occurring at the time, which saw the increasing social engagement of women, particularly within the fields of humanitarian and other charitable work. Her building was a place for 'lost' dogs and it was the largest building of its kind in the world at the time. Within years of opening, it would play home to as many as 12,500 dogs at once – some (but by no means all) rehomed among London's newly brimming bourgeoisie. Dickens, predictably, was a fan and in his magazine, *All the Year Round*, he referred to the new institution as an 'extraordinary monument of the remarkable affection with which the English people regard the race of dogs'. 'It is the kind of institution,' he wrote, 'which a very sensitive person who has suffered acutely from witnessing the misery of a starving animal would wish for, without imagining for a moment that it would ever seriously exist. It does seriously exist, though.'

Tealby was just one of many important figures to come. Another was Frances Power Cobb, anti-vivisection activist and women's suffrage campaigner. Then there was Lizzy Lind, the powerful orator and author of a tell-all exposé on the lives of laboratory animals. Each of these powerful, committed women would shape the science of the next century and inspire and energise fractioned sections of society in so doing. Their stories are told in upcoming chapters.

Dogs weaved a magic in those middle decades of the nineteenth century. This period saw dogs achieve something no other animal on Earth had managed. Somehow, in a matter of decades, dogs had captured our hearts. They had found their way into our homes. They had captivated our literature and, in time, they would steal away with our science. Through Darwin, through Dickens, through Tealby, through time … the dogs, against all odds, were both taming and being tamed. It is in this tempestuous era, in these turbulent decades, that our story of dogs in experimental science really begins.

CHAPTER 2

Emancipation Day

> 'In the agony of death a dog has been known to caress his master and everyone has heard of the dog suffering under vivisection who licked the hand of the operator; this man, unless the operation was fully justified by an increase of our knowledge, or unless he had a heart of stone, must have felt remorse to the last hour of his life.'
>
> – Charles Darwin (1871)

In the 1860s to the 1880s, Darwin's *On the Origin of Species* caused something of a boom in the status of science across Britain. Paid careers blossomed across scientific sectors – in museums, botanical gardens and field stations; by schools, teachers and by amateur naturalists operating from backyard sheds. As if a new Enlightenment had occurred, biology was unlocked. Freed, as if given a shot in the arm.

Equipped with a new approach for understanding animals, and seeing how effortlessly Darwin had deduced so much by studying animals so closely, many British naturalists in the last decades of the nineteenth century were directly inspired to investigate the animals nearest and dearest to them in life. In this chapter, we begin with three such naturalists. Each of these men would undertake the world's first investigations into the minds of dogs in the period between 1880 and 1895. In no small way, they were Darwin's disciples. As far as scientific rigour goes, one can think of Morgan, Lubbock and Romanes as the good, the bad and the ugly of British dog science. Their styles aside, each one saw in dogs an opportunity to test what their perceptions of the world may be, drawing on the new perspective that Darwinian thought had opened up.

First, the good. His name was Conwy Lloyd Morgan and, in the context of dogs, his work took place in 1890 or thereabouts, during preparation of his tome *Animal Life and Intelligence.* Morgan was a rather suave-looking man with a beard like Neptune that gave him a kind of commanding presence mixed with what his peers described as a 'sympathetic friendliness'. One of his big interests in life was how the science of the mind – thoughts, ideas, language – could be linked up with our physical understanding of it. He is said to have been inspired by his grandfather in this philosophical quest, who declared one evening at the dinner table, like some sort of Zen healer, that if a man did not experience sausages he could never truly know that there could be such a thing as a sausage. A young Morgan apparently listened quietly to the debate about cooked meat that followed, rapt. From that moment, his career choice was set in stone. Formally, he became interested in so-called 'mental evolution', a subject that incorporated both animal intelligence and instinct. Morgan was fascinated by the philosophical distinction between *why* things happen and *what it is* that happens. The first (*why*) is subjective and therefore outside the realms of science; the second (*what*) is firmly within it. Using one's eyes to follow the movement of a billiard ball rolling across a table, for instance, requires both anatomy and physiology – a physical body, in other words – but the process is also a mental one, given that there is a conscious entity watching it. This apparent duality of experience engrossed Morgan and drove him forwards.

Morgan trained under Huxley (he of 'Darwin's bulldog' fame) and was clearly inspired by many of Darwin's findings. Perhaps more than anyone at the time, Morgan wanted to know what was going on in the minds of animals and so, naturally, his interests turned to the animals most accessible at the time: dogs.

He was particularly curious about 'trial-and-error' learning in dogs – how the process of scratching and nuzzling and sniffing at things can unlock a potential solution or reward. Though Morgan joined others in expressing amazement at the behaviours of dogs, he saw these events as accountable by observable periods of exploratory behaviours beforehand. His most commonly cited example of trial and error involved his dog, Tony. It was said that Tony could, at will, nudge the latch on his gate to escape the confines of his yard. To passers-by, this behaviour suggested Tony was a deep conceptual thinker, capable of understanding the mechanics of hook and latch, but Morgan's stringent observations of the episode suggested that Tony's success was the result of a drawn-out period of trial-and-error learning, where the dog snuffled, nudged and pulled on the latch, only to stumble upon a solution accidentally. After this point, the gate-latch trick stuck in Tony; he was replaying solutions rather than reinventing them, Morgan noticed. Based on observations like these, Morgan would see Tony as a good dog, but a genius Tony was not. By way of an extra example of Tony's cognitive skills, Morgan spent days teaching him to turn his head while carrying a stick in his jaws so that both dog and stick could pass through the railings of a fence unhindered. Tony could not. We see the same thing today, of course; this geometric concept continues to be a stretch too far for many dogs.

Morgan's great legacy is the scientific rule that would forever come to bear his name – a simple test that animal behaviourists know best as 'Morgan's Canon'. This Canon says humans should only label animals as having emotions, intentions or understanding if less obvious explanations for these phenomena cannot be determined. Though the application of this rule may sound debasing to animals, this wasn't Morgan's intention; instead the Canon sought to avoid mislabelling behaviours unnecessarily. Morgan was applying a proof, in other words, to avoid anthropomorphism.*

Some might argue that our next scientist – the bad – could have done with applying Morgan's Canon a little more judiciously. John Lubbock was something of a polymath who dabbled in many, many things, so his occasionally scattergun approach to science is forgivable. He was certainly a man far closer to Darwin's heart than Morgan. In fact, living next door, young John Lubbock frequently played with the Darwins' children. Lubbock also shared Darwin's love of insects – so much so that it is said the two had matching microscopes. They really were close friends. At one point, in the midst of a depressive episode, Lubbock was the only visitor that Darwin would entertain.

* Morgan's Canon remains as important today as it was in Victorian times. The primatologist Frans de Waal calls it 'perhaps the most quoted statement in all of psychology'.

Though a career in banking would take up much of his time, it was Lubbock's private passions for which he is most fondly remembered: his discovery of numerous mite species, his studies of invertebrate nervous systems; his coining of archaeological terms, like 'Palaeolithic', 'Mesolithic' and 'Neolithic'. He was even one of the first to use the term 'prehistoric'. He would later, in a successful stint as a politician, go on to invent the concept of the bank holiday.

Lubbock was also one of the first individuals to use experiments to assess the behavioural responses of dogs. Like Morgan, his own dog (a poodle named Van) was used as his experimental subject. It took more than three months of training but, on the surface of it, what Lubbock achieved with Van was a feat approaching the Doolittlian. In the most basic interpretation of events, Lubbock taught Van sign language. And I mean that literally – with little signs. If you want to replicate Lubbock's experimental set-up at home you will need to arrange some tiny placards on which you can write words. First, write on one card, in big letters, the word 'FOOD' and then call your dog over. When your dog happens to pick up this card with its jaws, run off and get them a treat. In only a few hours or so of this training regime, you'll notice a change in your dog. The dog will start picking up the 'FOOD' sign regularly. You'll know things are going well when you find yourself being worked by your dog like a vending-machine claw, running back and forth to acquire treats from the kitchen morning, noon and night. You can introduce other signs at this point. For instance, you can write the word 'WATER' in large, bold letters and change the reward for a bowl of water. Over time, the same pattern will emerge in the dog's behaviour. It now apparently knows two words: 'FOOD' and 'WATER'. Congratulations, you are training the dog and, in return, it is training you. A relationship will begin to blossom between them, the signposter, and you, the table-waiter.

A month or so into Lubbock's working with Van on this simple procedure, he decided to add two new words: 'OUT' (as in 'open the door') and 'BONE' (as in 'give me a bone'). Van duly took to the challenge, coming to wield each expertly. 'No one who has seen him look along a row of cards, and select the right one, can, I think, doubt that in bringing a card he feels that he is making a request, and that he can not only perfectly distinguish between one word and another, but also associates the word and the object,' Lubbock surmised. He even spelled out some cards phonetically to help his little canine student get the hang of the English language. Morgan, no doubt, would have spat feathers at this bit.

Though the idea is easy to ridicule, Lubbock was making some sense. After all, the approach is not wholly dissimilar to chimpanzees and dolphins who, in the modern day, can be taught to press buttons and touch-screens for their desired objects or rewards. Yet Lubbock's study left the door open to misinterpretation, giving some people the idea that dogs were using words rather than picking up on the patterns of wavy lines and smudges on each sign to get what they wanted – something Morgan's Canon would suggest was the simpler of the two interpretations.

In part, Morgan devised his Canon in direct response to another animal scientist who was making waves at the time. That scientist was George Romanes, author of a number of notable works including *Animal Intelligence* (1882), a rival to his own book on animal intelligence and perhaps the first book about comparative psychology. Romanes is the third of our early dog scientists, and it was between Romanes and Morgan that things got ugly.

Though they differed in age, Romanes and Darwin were – like Lubbock and Darwin – firm friends. Before becoming an influential writer about animals, Romanes was Darwin's research assistant and confidant. He was his pen pal, and some scholars suggest Darwin was a father figure to Romanes. Darwin's literary style – at once conversational and quick to include anecdotes – rubbed off on Romanes. In fact, before his disappearance from the history books, many considered Romanes might be Darwin's rightful heir.

Over their years together, the two shared many interests – one, predictably, was dogs. Romanes was so devoted to dogs, he gave a whole chapter to them in his 1883 treatise, *Animal Intelligence*. Both men considered dogs capable of a deep capacity for emotion, seeing in them 'pride, sense of dignity, and self-respect'. However, in an interesting sign of the times perhaps, Romanes considered only the most well-nurtured dogs to be capable of eliciting such intelligence. Mongrels ('curs', as they were commonly called then) had no chance.*

Like Darwin, Romanes produced a highly readable account of animal intelligence but, at the time, his approach was not considered by all to be particularly scientific. Yet he was first to undertake a formal scientific experiment to understand the senses of dogs. In 1887, he tested the scent-detecting abilities of his setter, with whom he had hunted for eight years.† In the experiment, Romanes and his friends gathered in a local park and made plans to walk around indeterminately. In the first test, the setter would be encouraged to follow Romanes' path along the field, a challenge the dog duly met without apparent problems. Then Romanes undertook another rambling path through the park, but this time he and his friends removed their boots. In this test, the setter struggled to find the scent. His experimental subject was following the scent of footwear, deduced Romanes. To test if this was indeed the case, he and

---

* The word 'cur' is thought to come from the Old Norse *kurra*, meaning 'to grumble or growl'.

† Very little exists about the name of this dog but, looking through Romanes' letters, it could well be Flora, a setter bitch he refers to as 'a beauty', like his former dog Bango 'but with a prettier face'.

his friends swapped boots. The dog followed his shoes, and his shoes only. His conclusion was simple: had he been 'accustomed to shoot without boots or stockings, she would have learnt to associate with me a trail made by my bare feet'. This was quite a significant leap: though many suspected dogs depended heavily on scent for their movements, Romanes was the first to prove it.

In Romanes' writing and in his public lectures, his feelings about animals rang loud and clear. He saw intellectually gifted animals like dogs as being able to process abstract ideas gained from their sensory experience of the world. He claimed that animals were capable of experiencing 'all the human emotions except those which refer to religion and to the sublime'. He saw in us the role of teachers – that humans could help endow pets, particularly dogs, with moral senses. Naturally, the hackles of his detractors were raised at this. One of the fiercest was Morgan, who argued 'one should, in such a situation, attribute as little intelligence as their acts would justify.' Morgan was particularly opposed to Romanes' occasional dips into 'anthropopsychism' – his attribution of consciousness to divine beings and nature. In his own magnum opus, he refers briefly to Romanes' 'adequate knowledge and training' before delivering blow after blow of withering scepticism about his over-reliance on anecdotes. For Morgan, science was about reasoning above all else. Yes, animals were intelligent, he surmised, but only 'man' had a faculty for ideas, ideals and morals – a capacity for reasoning, in other words.

Though they may have disagreed on their experimental methods, Romanes, Lubbock and Morgan clearly saw potential in dogs; each leaned heavily on their personal experiences with dogs, letting their dogs lead the way towards a more proper scientific understanding of their study animals. Their arguments were shaped and chiselled by the close relationships they shared. Yet, to a degree, in the scientific circles of Victorian Britain, Darwin and these dog disciples – questioners, wonderers, sharers of all information, no matter how insignificant – were a dying breed. Rather than working in front of packed balconies of research students, eager to see dissections on animals both living and dead, Darwin and his disciples spent their time considering nature from the warm confines of a well-heated study, surrounded by books, with a dog asleep at their feet. This 'gentlemanly' approach to science began to date. Increasingly, science was a practice that required a more controlled arena. A clean space where variables could be controlled and observations isolated from the noise that existed in nature.

'Scientists were a new kind of priesthood, an elect, who insisted on the primacy of their own values against the soft morality of ordinary people,' writes John Homans in *What's a Dog For?* The problem was that soft morality is still morality. And morality, to paraphrase Mark Twain, has a habit of gratifying half of the people while astounding and enraging the other.

As tensions continued to mount between rival factions of science, they rubbed up against wider insecurity about the role of science in society, particularly when it came to dogs. Rabies did very little to help matters. The pot began to boil …

Even with the continued removal of street dogs, cases of human deaths from rabies remained on the rise in the final two decades of the nineteenth century. In the period between 1884 and 1885, for instance, human cases of rabies in the UK doubled and more than a thousand dogs were known to have had the disease. Many of these rabies cases were in the capital, some no doubt just a few hundred metres from Parliament. Unsurprisingly, with pressure from the public mounting, politicians began to double-down their efforts to curb the disease – indeed, there was hope that one day a vaccine might be invented. Alas, there were no details of when it might arrive and who might make it and so, to quell panic, the government and local officials took matters into their own hands, ordering that all pet dogs should wear muzzles when outdoors so as to reduce any further potential spread. Outcry from some corners of the dog-owning community was inevitable.

Many of the arguments about muzzles that raged at the time will be recognisable from our own era: the Victorian debate saw a medical community united in prioritising healthcare and minimising disease risk, pitted against private individuals who saw the rights of their dogs slowly being chiselled away. Muzzles and other restraints made barbarians of loyal pets, they argued. 'The muzzled dog is a dog constantly tormented and oppressed,' wrote Ouida, the pen name of English author Marie Louise de la Ramée. 'For public health doctors, dogs were these dangerous bodies which had to be disciplined for the good of the public's health,' says Professor Abigail Woods, animal historian at the University of Lincoln, UK. 'But for their owners, dogs were family members whose compulsory muzzling by government diktat amounted to unjustifiable state intervention.'

As is the wont of Victorians, societies were soon set up to lobby government on the matter of muzzling. On the side of muzzles was the Society for the Prevention of Hydrophobia and the Reform of the Dog Laws (SPH), formed in September 1886 – within a month of the formation of its rival society, the anti-muzzling Dog Owners' Protection Association (DOPA). The latter society was formed as a response to the 'Baker Street affair', in which a lady was arrested after pouring water over the head of a police inspector who had given orders to have her unmuzzled dog put to death on her own front doorstep. For more than a decade, the two societies lobbied, agitated and generally campaigned about their cause to all who would listen.

To call it a pressure cooker atmosphere would be to downplay the stressors: first, a rampant and highly contagious disease; then, removal of street dogs, the key vectors of transmission; then, with nothing helping much, government interventions to muzzle pets. The pot was really bubbling now.

The arrival of the rabies vaccine, when it was finally produced, did almost nothing to reduce public anger and resentment of the government's handling of the situation. Why? Mostly because the treatment required those suspected of being bitten by a rabid dog be shipped off to Paris, which was time-consuming and costly. But also because the long-awaited treatment came from none other than Louis Pasteur, a scientist many considered to be a serial abuser of animals, particularly dogs, a species upon which many of his experiments were undertaken.* 'Pasteur's reliance on vivisection for his breakthroughs simply made things worse, and anti-vivisectionists and dog defenders refused to accept either the moral or the scientific authority of "Pasteurism",' writes Philip Howell in the illuminating *At Home and Astray*. To the anti-vivisection movement, with Frances Power Cobb at its charge, this was the final straw.

A skilled journalist and something of a reformer, Frances Power Cobbe became so incensed about what she discovered was going on behind the doors of medical laboratories that she devoted her wealth, her politics and every last ounce of fight to challenge it. In 1875, she founded the Society for the Protection

---

* Working alongside the laboratory's co-founder, the bacteriologist Emile Roux, Pasteur went to extraordinary lengths to get the rabies vaccine tested. Not only did this involve harvesting rabies from the mouths of rabid dogs held down by men in leather gloves, it also involved trialling the vaccine on infected children who had been bitten by dogs and were desperate for treatment – something Pasteur had no licence to do. Thankfully, the first proper trial worked out: nine-year-old Joseph Meister was saved by the vaccine on 6 July 1885.

of Animals Liable to Vivisection (SPALV) and later she would found the British Union for the Abolition of Vivisection (BUVA), both of which are still active today. Cobbe was able to tune into, and unify, the national frustration at the way in which the medical sciences used dogs for their research. But there was a certain nationalism there too. At the time, the British were especially sniffy about the French, for instance, who were leading the way in the exciting new science of physiology, duly demonstrating their discoveries to visiting researchers, often using dogs as their subjects. According to Robert Kirk's wonderfully insightful chapter in *The Routledge Companion to Animal–Human History*, Cobbe regularly aimed her frustrations across the Channel: 'Cobbe frequently characterised vivisection as a French corruption of science arguing "as a rule that the most cultivated are the most merciful" yet in "France, alas! It is men of science – men belonging to the learned professions – who disembowel living horses and open the brains of dogs."'

Only the year before the foundation of SPALV, the British had been particularly enraged at a public demonstration (performed in Norfolk) by one French anatomist that saw a dog tethered up and, without anaesthetic, being injected with pure alcohol or creosote to show how poisons act on the central nervous system. 'The animal struggled much, cried as far as it was able, showed other symptoms of great suffering, and ultimately – not long after the injection – had a fit of epilepsy,' reported the *British Medical Journal* at the time. Before being sentenced to appear in court over this apparently illegal act, the French anatomist in question fled back across the Channel, much to the disappointment of the anti-vivisectionists.

The case was particularly charged since it reminded anti-vivisection campaigners of a former bête noire of their movement, François Magendie (1783–1855). To modern audiences, Magendie sounds nothing short of monstrous, almost like something dreamed up in a Hammer Horror movie, so it is no surprise there was a real and genuine hatred of him from many quarters. Some argue he started the whole anti-vivisection movement off when he performed his gruesome demonstrations at London's Windmill Street anatomy school in 1824.* One typical account describes Magendie's style: 'Monsieur M. has not only lost all feeling for the victims he tortures, but he really likes his business. When the animal squeaks a little, the operator grins; when loud screams are uttered, he sometimes laughs outright. The professor has a most mild, gentle and amiable expression of countenance, and is in the habit of smoothing, fondling and patting his victim whilst occupied with preliminary remarks ...'

As harrowing as these encounters read, it is worth underlining that many, if not all, practitioners of vivisection saw themselves as doing moral good in society. In fact, many saw themselves as something approaching the humanistic in their spiritual leanings. The renowned French physiologist and vivisector Claude Bernard (1813–78), the man who paved the way for the discovery of how animals maintain body temperature through physiological processes, regularly espoused this viewpoint. Like others, he saw animals as 'living machines' and believed that 'the science of life can be established only by experiment, and we can save living beings from death only by sacrificing others.'†

Not everyone saw it this way, of course. Most notably, his wife (Marie Françoise Martin) chose to leave him over this conflict of interest and became a leading figure in the

* The details I will spare you, but know that the first lecture of each spring term involved Magendie's students turning up to a laboratory in which there was a basketful of live rabbits, eight glasses of frogs, an owl, two pigeons, some tortoises and a puppy. The rest ... well, you only need imagine.

† To understand the basics of what would later be known as 'homeostasis', Bernard essentially cooked animals, while still alive, in a makeshift oven he had installed in the basement of his laboratory.

anti-vivisectionist movements bubbling up in polite society across Europe. Yet Bernard's research continued, apparently unfazed: 'The physiologist is no ordinary man,' he wrote. 'He is a learned man, a man possessed and absorbed by a scientific idea. He does not hear the animals' cries of pain … He sees nothing but his idea, and organisms which conceal from him the secrets he is resolved to discover.' It was inevitable that, by the latter stages of the nineteenth century, tensions between the rival camps would begin to boil over.

The good news was that, in the end, the rabies crisis did recede in London and in most other Western nations. It may have been because of the muzzles. It may have been the removal of so many street dogs. In truth, it was probably a bit of both. There are stories of dog owners celebrating in their local parks when rabies subsided completely. Of dogs running wild and free, their muzzles dangling from their faces. Of dogs dragging home-made banners attached to their collars that celebrated 5 January 1891 – 'Emancipation Day', as one lively dog owner called it. In truth, dogs' emancipation from the shackles of science was still a long way off because, hot on the heels of Pasteur, was a man named Pavlov, and his treatment of dogs would be no less polarising.

CHAPTER 3

Sacrificed for science

> 'To tell us that every species of thing is endowed with an occult specific quality by which it acts and produces manifest effects, is to tell us nothing; but to derive two or three general principles of motion from phenomena, and afterwards to tell us how the properties and actions of all corporeal things follow from those manifest principles, would be a very great step.'
>
> – Isaac Newton, *Optics*

In a sense, we are all Pavlov. Whenever we go to the kitchen cupboard and grab the dog food and then place the dog food on the counter, the noises we make – the specific clinking and clanking of the dog's bowl – is changing our dogs' universe in an easily overlooked, but no less important, way. First, the bowl hits the kitchen counter, creating a Mexican wave of vibrating atoms that radiate outwards into the universe – a sound wave. The sound wave travels in every direction, some moving into your dog's ears, where it then tickles specific auditory cells. These cells discharge a tiny electrical impulse that travels like lightning into the dog's brain. The exact pattern of electrical signals is registered. It correlates with those previously received by the brain. An unthinking recognition occurs and a new electrical impulse is elicited. This one travels from the brain and heads to your dog's salivary glands. Your dog looks up expectantly. Juices begin to flow from these chemical factories. In the mouth, digestion begins. Congratulations, you have classically conditioned another life form on planet Earth. You have induced a Pavlovian response in your pet.

Ivan Pavlov (1849–1936) never set out to discover the phenomenon for which he would later become best known. Instead, his research interest was digestion, specifically the role of digestive juices produced by the body. In fact, it was for this research that he was awarded the Nobel Prize in 1904, by which point his laboratory at St Petersburg's Institute of Experimental Medicine had become a notable seat of influence in the field of physiology.

Pavlov's success was down to his insight, sure, but it was also because of his devotion to repeated experimentations in controlled conditions that could be thoroughly documented. If the nineteenth century was about fuzzy, armchair anecdotes, Pavlov was part of a new era – one of a more rigorous science. Of collection, ongoing studies, and long-term trials. As such, his laboratory was set up with factory-like precision. He saw his dogs as something akin to machines and their digestive juices as crucial products that could be collected, detailed, studied and measured. Where many animal experimentations involved dissection of animals that were subsequently dispatched and disposed of, Pavlov was aware that the workings of glands are a long-term process. Thus, he favoured an approach of 'chronic experiments' in his laboratory. This required that some dogs be kept alive for as long as possible while experiments were undertaken. The digestive juices he was most interested in were collected over a period not of minutes or hours but weeks or, ideally, more. Today, we think of such long experiments as barbaric, but this kind of research was becoming increasingly common in many laboratories of the time.

Here is a brief overview of how the process of collection worked. Those of a particularly sensitive disposition may prefer to skip the paragraph that follows.

Pavlov and his team surgically operated on dogs by making an opening in their digestive tracts – a fistula – through which the workings of the body's internal organs could be observed, in part, by the scientists' own eyes. This allowed Pavlov to observe and measure the amount of gastric juices being produced. It also meant that tubing inserted into the fistula could collect juices leaking out over days or weeks into a nearby pouch or jar.

Many dogs would die during the course of these experiments, as you might imagine. With little by way of anaesthesia or analgesics, all would suffer. But there is no doubt that the laboratory, by undertaking experiments like these, charted Pavlov on a course towards a rich shore of uncharted scientific territory.

Within years of his studies beginning in 1891, according to Daniel T. Todes' illuminating biography, Pavlov and coll-eagues produced reams of research data on the physiology of the digestive system and previously unknown gastric juices produced as a by-product. These gastric juices are worthy of a brief aside, because at one stage there were so many bottles of the stuff being produced that Pavlov saw an opportunity for a slightly ghoulish and lucrative side-gig.

Polite society had long been looking for a cure for a common ailment, that of dyspepsia (otherwise known as indigestion), caused by the body having trouble digesting certain foods. Dyspepsia causes nausea, discomfort and bloating and so, seeing a potential application, Pavlov took his moment. As well as undertaking experiments, he bottled up crates of the stuff harvested from recently fistulated dogs and sold it directly to sufferers, who lapped it up (so to speak) with abandon. By 1904, Pavlov was harvesting enough gastric juice to fill 3,000 bottles or more each year, resulting in profits that would increase the laboratories' coffers by more than 70 per cent. It was during this busy period that Pavlov noticed the strange phenomenon that would later take his name.

To really get the juices flowing in his dogs, Pavlov would place a bowl of minced meat close enough for his dogs to see and smell, but just out of reach. Seeing this bowl of delicious dinner was enough to stimulate the production of saliva and the gastric juices required for digestion. But sometimes, some of the dogs would start producing saliva before the appearance of food. It was like they had a sixth sense for when the food was going to show up. For this reason, Pavlov called saliva a 'psychic secretion'.

Further investigation proved that the dogs had keyed into tiny signals in the immediate environment that correlated with feeding time. The dogs would salivate upon seeing the dog's bowl being lifted up, for instance, or the sight and sound (or smell) of the same person appearing, at the same time, over a period of days. Pavlov's great insight was to spot that his dogs' response came about not through any kind of considered thought; he observed that the dogs were responding involuntarily. Pavlov realised that the brain had found a kind of behavioural short-cut in its production of saliva. One that by-passed active consideration of any kind. He and his team set to work to experimentally induce salivation in the dogs through artificial means.

Todes' findings, expertly outlined in *Ivan Pavlov: A Russian Life in Science*, are revealing. Pavlov 'never trained a dog to salivate to the sound of a bell … Indeed, the iconic bell would have proven totally useless to his real goal, which required precise control over the quality and duration of stimuli.' In fact, 'bell' was likely a mistranslation of the Russian *zvonok* meaning 'buzzer'. As well as buzzers, Pavlov used whistles, harmoniums, metronomes, tuning forks and – distressing as it sounds – electric shocks. In each case, the dogs appeared to have no problems accustoming (or 'conditioning') themselves to these novel stimuli.

In the years that followed his initial discovery, Pavlov took these experiments further. He investigated how fine-tuned the stimulus response could become, how sensitive the brain was to picking up on unrelated stimuli and how this changed with time. With metronomes, for instance, he undertook tests where the dogs only received their reward when the metronome ticked at, say, sixty beats per minute. At first, the dogs would salivate upon hearing any metronome setting. In time, by reinforcing the dogs' responses with food, at exactly the right setting Pavlov could explore how sensitive the brain was in determining reflex actions.

Pavlov was, in no uncertain terms, challenging what was then widely considered to be true in science: that brains were unexplorable black boxes that would forever be hidden from science. The mind was not a black box, his research suggested, but an observable, elegant machine open to scientific questioning like any other organ. He saw the application of his research as a means to open 'the mechanism and vital meaning of that which most occupied Man – our consciousness and its torments.' This really was a step-change in how we think about the workings of the brain. It is why Pavlov's ideas spread across the world, courtesy of the publishing of his findings in a book that underwent translation into many languages and was subsequently distributed around the world.

Called *Conditioned Reflexes: An Investigation of the Physiological Activity of the Cerebral Cortex*, it caused something of a smash, particularly in the newly cementing psychological circles of North America. In no uncertain terms, Pavlov's rigorous experimental set-up made him something of an academic god. In time, others would join him on his cloud.

One person Pavlov's discoveries particularly resonated with was the American psychologist John B. Watson (1878–1958), who would go on to found the field of Behaviourism. Watson's academic take was that, like Pavlov's experiments upon brains and glands, psychology could be broken down into scientifically observable parts that could give insight into human minds. He wanted psychology to move from the realm of philosophy and align itself with the biological sciences, in other words. From relatively humble beginnings, Watson's rise to power from his seat at John Hopkins University (Baltimore) saw his ego sky-rocket to almost Trumpian levels by the time he hit his stride. As he put it: 'Give me a dozen healthy infants, well-formed, and my own specified world to bring them up in and I'll guarantee to take anyone at random and train him to become any type of specialist I might select – doctor, lawyer, artist, merchant-chief, and, yes, even beggar-man and thief, regardless of his talents, penchants, tendencies, abilities, vocations, and race of his ancestors.'

Where Pavlov most famously worked with dogs, Watson famously worked with human babies. Or rather, he worked with one: a nine-month-old baby named Albert, known more colloquially in psychological circles as 'Little Albert'. According to notes, Watson is said to have rented the boy from his mother, a wetnurse, for the princely sum of one dollar. The return on the investment for research circles was very good. For the mother, less so. The test itself became very important for the fledgling field of Behaviourism as a kind of proof of concept. It was also a rather neat example of how wayward psychological research could be before the invention of ethics boards.

Directly inspired by Pavlov, Watson began his experiment by introducing to Albert a white rat, which initially elicited in the baby a warm and happy response, though this reaction would not last long. After a few reintroductions of the rat, Watson set to work on his plan: upon getting out the white rat for Albert to see, Watson began to introduce the surprise sound of a hammer being struck with force against a metal sheet. The resultant loud clang clearly scared Albert and, unsurprisingly, the noise elicited a fearful, tear-filled response. Watson repeated the set-up. There were tears. Then more tears. Then more. Predictably, within a few repetitions, Watson had what he wanted: a baby that spontaneously and uncontrollably demonstrated fear at the sight of a white rat.

Clearly, in the modern day, ethics boards would baulk at such an experimental approach – and frankly, who could blame them? – but the experiment threw up some interesting side effects. One was that the brain, in response to the visual stimuli of a white rat, seemed to be picking up on the concept of 'fur' rather than 'rat'. After a few repetitions, Albert began to demonstrate a fear response not only upon seeing white rats but also at seeing furry things more generally – stuffed toys, rabbits, fur coats, dogs and (rather perversely) a white Santa Claus beard that Watson must have had lying around. Watson's conclusion was that emotions could be classically conditioned in a way that was similar to the salivary glands of dogs. Emotional responses could be elicited by external stimuli, in other words. To put it simply: experiences could influence emotions – and, importantly, they could be observed, manipulated and studied. He concluded that Pavlov's ideas really could be applied to psychology.

And what of Albert, you might ask? Watson's plan was to recondition Albert's brain at the end of the experiment so that he could behave normally upon seeing white rats (and Santa) in the future, but sadly this wasn't to be. Before he had his chance, Albert's mother (clearly very upset at Albert's treatment) removed him from the experiment prematurely.* Instead, the great prize for doing the first ever 'unconditioning' of a patient went to a different researcher: the psychologist Mary Cover Jones (1897–1987).

Having sat in a lecture theatre and heard Watson speak about Little Albert, Jones went off and devised a treatment for a two-year-old boy named Peter who had developed a fear of rabbits. The therapy, though simple, was highly effective. Again, it drew from Pavlov's notion of conditioning. To counteract the boy's rabbit fear, each day Peter would be encouraged to sit closer and closer still to a caged rabbit while he ate his lunch until, eventually, the rabbit was close enough to touch. In addition, Peter was regularly shown his friends interacting with rabbits and enjoying themselves. In time, the two actions appeared to change Peter's nervous response. To rapturous applause, Peter was cured and Mary Cover Jones became the first psychologist to 'uncondition' a fear response. For this and other discoveries, she is respectfully known as 'the mother of behaviour therapy'.

Watson and Jones were just some of the people taking Pavlov's ideas and applying them in new places. Another was Edward Thorndike, the American psychologist (1874–1949) and the coiner of psychology's Law of Effect. Thorndike's primary interest, rather like those of Darwin's disciples before him, was investigating how animals solved problems. He wanted to discover a way to plot the mechanisms involved in animal learning. For this, he used specially prepared puzzle

---

* What happened to 'Little Albert' was one of psychology's great mysteries until a 2012 article, reporting on researching into historical hospital records suggested, sadly, that he died of acquired hydrocephalus at just six years old. Other researchers have contested that another man, Albert Barger (who died in 2008) was the mysterious baby. Interestingly, Barger was reported to have had a lifelong fear of animals.

boxes that were fundamentally little wooden cages with ropes, pulleys and sliding doors – animal escape rooms, if you will. To escape from a puzzle box, Thorndike's animal subjects (which included both dogs and cats) were required to press on a specific lever or pull a cord or slide a latch using a paw. Once the action was successfully undertaken – hey presto! – the research animal could escape and receive its food reward. But the test didn't end here. After this first learning experience, the animal subjects were put back in the box to see how quickly they could escape a second time. And then a third time. Then a fourth. Through these puzzle boxes, Thorndike's experiments aimed to challenge the assumption that cats and dogs solve problems like we do: through careful consideration of the information available, including by drawing on experience – thinking it out. Thorndike was essentially looking for insight, but he failed to find it in his cats and dogs. Instead, his results suggested that, upon re-entering the puzzle box, the cats or dogs would scrabble around in much the same manner as in the first trial, using trial and error to hit upon solutions, finding success only slightly more quickly each time.

What was particularly exciting – ground-breaking even – and what set Thorndike apart from Darwin's disciples, was that the results of his puzzle-box experiments could be plotted on a graph, with trial numbers along the *x*-axis and the time it took to escape along the *y*-axis. This allowed him to chart each animal's experience in the puzzle box, from the first disorganised and rather chaotic attempts at escape, to several goes in when the cats and dogs were starting to establish the routines or techniques that would result in their being released and rewarded. Plotted on a graph, Thorndike was an early practitioner of 'learning curves', where the time it took to escape (or 'performance') was placed along the *y*-axis and the number of trials ('experience') placed along the *x*-axis. As the subjects gained more experience of the experimental set-up, their performances improved until, after thirty trials or so, they managed their escapes within seconds (rather than minutes) each time.

There was another early practitioner of learning curves who deserves equal place in this book – a zoologist who is almost certain to be missing from most psychology or zoology textbooks but whose insight and flair for experimental design would have led our story down a completely different path, had he been recognised by his contemporaries. His name was Charles H. Turner (1867–1923). Though Turner published more than seventy papers, his impact on early animal intelligence studies has been almost completely overlooked, in large part because of insurmountable barriers caused by his African-American ethnicity. His research animals included birds and invertebrates, particularly spiders.

Turner was especially interested in an important component of Darwin's theory – that of variation. To understand variations in behaviour between individuals, Turner turned to spiders and how they produce their webs. He observed how webs differ according to the geometric space available to the spider: 'we may safely conclude that an instinctive impulse prompts gallery spiders to weave gallery webs, but the details of the construction are the products of intelligent action.' By undertaking experiments with bees, ants and wasps, Turner quietly observed feats not only of intelligence but also of temperament – personality, essentially. His studies of cockroaches were particularly fascinating. By putting cockroaches in mazes, Turner noticed that older cockroaches took longer deciding over their routes, yet achieved more precise results. He saw the stationary deliberation that these older cockroaches went through at various points in the maze as evidence of some sort of consciousness.

The paths of Thorndike and Turner did occasionally cross. Where Thorndike favoured the idea that all animals preferred trial-and-error learning, Turner did not. In 1912, he confronted Thorndike with evidence garnered from his personal experience of watching wasps walk across the ground while carrying heavy prey back to their nest. Turner saw that these wasps could navigate numerous obstacles on their journey in a way that was purposeful and in no way blundersome, as Thorndike's notion would have predicted. He also cited his observation of how ants isolated on tiny islands begin using nearby materials to try and build little bridges back to 'the mainland'; again, such behaviour implied a deeper mechanism – 'thinking' or something like it.

'Even though his experimental work was known to contemporary giants such as John Watson and Thorndike,' write modern-day biologists Samadi Galpayage and Lars Chittka, 'Turner's visionary ideas about animal intelligence did not resonate in the field; perhaps they were simply too far ahead of the time. Accordingly, they are almost completely unrecognized in the current literature.' Their research paper quotes the African-American historian, William Du Bois (1868–1963): 'C. H. Turner, one of the great world authorities on insects, nearly entered the faculty of Chicago university; but the head professor who called him died, and his successor would not have a "N—."' Had Turner had access to the facilities and resources of his contemporaries – had his skin been a different colour – the whole history of ethology, cognitive science and psychology may well have turned out very differently. In fact, this book would chart a completely different path. Science prides itself on being rational. It is pure in application. But not all parts of society are free or able to apply themselves – this applied as much then as it does now. The result is that Thorndike's legacy lives on and Turner's does not.*

---

* In the Research Notes at the end of this book, Samadi Galpayage has been kind enough to offer me some advice for readers who are keen to learn how they might rediscover and celebrate other pioneers of science whose influence has long been overlooked.

In addition to the learning curve, Thorndike also gave us his famous 'Law of Effect' (1898), which states that behaviours that give satisfying outcomes tend to be repeated (or *strengthened*) and those that provide undesirable outcomes tend not to be repeated (they become *weakened*). This would become a key piece of the behaviourist jigsaw and, alongside Watson and Pavlov's ideas about conditioning, would influence the field of psychology immensely in the decades that followed.

Through the use of an elaborate puzzle box, Thorndike also stumbled across the idea of 'operant conditioning' – that individuals learn to make a clear association between a particular behaviour and a consequence. At its simplest, operant conditioning is when your dog sits on demand, in expectation of a treat.*

Watson, Thorndike, Jones – in a way, these scientists and psychologists have become a part of Pavlov's legacy, which opened up the black box of the mind, allowing humans to see a sliver of light in what may or may not be going on. A way, perhaps, to study the mechanisms of brains, decision-making and emotions for the first time. They have Pavlov – and his dogs – to thank for it.

As we reach the end of this chapter, I confess a strange kind of sadness that we know so little about these dogs. I am embarrassed to admit that, many years ago, when learning of Pavlov's studies in the early years of my studies, I swallowed a viewpoint of Pavlov that is classically human-centric and failed to give any consideration of how these dogs lived and how they must have suffered. I now realise that this wasn't all my fault. That, to a degree, my education misled me. I think

---

* Operant conditioning differs from classical conditioning in that there is an element of choice in the matter. Dogs cannot choose not to salivate, for instance. However, they can choose not to sit down or roll in faeces or chew on socks.

education is still like this. A quick search online for learning resources about Pavlov and his discovery of classical conditioning shows a surprising number of slides and printouts of cartoon dogs, pictured with smiles, sitting respectfully in front of clip-art renderings of bells, with arrows pointing out the gathering saliva. Some of these learning resources are from very respected universities. I remember images on slides just like these when I was at university. I think this is why I never thought much more about the dogs. Perhaps I was, myself, conditioned. The truth is that their untold suffering – and it really was nothing less than that – was, in no small way, a fledgling science's gain. And so it is right that we remember those dogs whose names we know, even if we know little about their characters, temperaments and experiences. Especially if it helps us vow never again to make such easy slaves of their desperate interest in connection.

For posterity, here are some of their names: Arap, Arleekin, Avgust, Baikal, Barbus, Bek, Bes, Bierka, Box, Boy, Chernukha, Chingis Kahn, Chyorny, Diana, Druzhok, Felix, Garsik, Golovan, Ikar, Iks, Jack, John, Joy, Jurka, Kal'm, Kellomäki, Khizhin, Krasavietz, Lada, Laska, Lis, Lyadi, Mampus, Martik, Max, Mikah, Milord, Mirta, Moladietz, Murashka, Nalyot, Nord, Norka, Novichok, Pastrel, Pestryi, Pingiel, Rex, Rijiy I, Rijiy II, Rogdi, Ruslan, Satyr, Serko, Shalun, Sokol, Sultan, Tom, Toy, Trezor, Tungus, Tygan, Rosa, Umnitza, Valiet, Visgun, Zheltyi, Zhuchka, Zloday, Zmei and Zolotistyi.

CHAPTER 4

The Brown Dog Affair

> 'The question is not, Can they reason?, nor Can they talk? but, Can they suffer?'
>
> – Jeremy Bentham (1789)

By the time of Darwin's death in 1882, evolutionary biologists (who did not yet know of Pavlov's work) found themselves in a kind of academic fug about what brains were and how they went about their apparent magic. On the one side, Romanes had proclaimed that all matter was somehow conscious (panpsychism) and that studying and understanding how the mind works was somehow outside of the realms of possibility. Then there was the other side, proclaimed by Lloyd Morgan, that behaviours and experiences very much *were* a product of natural selection and that we just needed to ask better questions of animals than scientists (like Romanes) were currently asking. But there were other viewpoints. Thomas Henry Huxley, for instance, was gaining some notoriety for his position that animals were something like 'conscious machines' or 'automata' – that consciousness somehow springs up as a 'collateral product' of the natural workings of animal bodies or, as he put it, 'the steam-whistle which accompanies the work of a locomotive engine is without influence upon its machinery'.

Of these two viewpoints, Alfred Russel Wallace – co-discoverer of natural selection – favoured Romanes' conclusion, just with added ghosts.* This came as no surprise to many at the time, of course. Within years of Darwin publishing *On the Origin of Species*, Wallace's interests had moved into spiritualism, following an incident with a floating lady, some cut flowers and a dash of ectoplasm.

According to Matthew Cobb's marvellous *The Idea of the Brain*, many of Romanes' ideas were closer to those of his contemporary French philosophers, who argued that thoughts were made of something unseen and immaterial, something unknowable and unobtainable: 'If thought were in the head, it would occupy a place there, which by dissecting one could end up finding on the end of a scalpel,' argued Henri Bergson (1859–1941), one of France's most influential thinkers. That's not to say that scalpels didn't have their uses, of course. The anatomists and physiologists, in particular, were making great

* Reading through Alfred Russel Wallace's letters, it is hard not to feel a little sorry for him. Not only would future generations come to remember him as a kind of supporting actor in the Darwin story, but he was also the victim of a tragic shipwreck event in which his whole zoological collection was lost. To make matters worse, within months of finding his way back to dry land, he was jilted by his bride-to-be at the aisle.

advances in their sciences in these exciting decades. And a post-Industrial world was giving them plenty of tools to play with.

Engineering, in particular, gave scientists a new way to consider the workings of the animal world, particularly the brain. The French anatomist Mathias-Marie Duval (1844–1907) saw brains as control rooms with switches, for instance. Likewise, French anatomist Louis-Antoine Ranvier (1835–1922) drew upon undersea telegraph cables, and their insulated covering, when considering the reason for the thick covering (called the myelin sheath) seen on motor and sensory nerve cells in vertebrates. Many scientists of the age, with one eye on the great strides being made in communications technology, came to consider the brain as an enormous telephone network, full of switchboards interacting busily. Yet there were some who continued to draw upon the ideas of French philosopher René Descartes (1596–1650) – the most famous exponent of the idea that minds were machines, powered by (in his view) roaming spirits that circulated throughout animal nervous systems, moved around by hydraulic pressure.

For much of this research, dogs – being fairly easy to get hold of, easy to train and easy to keep – continued to be useful, if not vital, research vessels. Without going into too much gory detail, many of these questions saw dogs at the literal sharp end as scientists searched desperately for a clearer understanding of how brains worked. By stimulating with electrodes certain parts of a dog's brain, for instance, German physicians discovered in 1870 that their subjects involuntarily moved certain parts of the body. This helped pave the way for our modern interpretation of the brain as a leading seat of power in the workings of the body. And, crucially, it gave scientists evidence that some regions of the brain are localised, each part responsible for eliciting reactions in particular parts of the body. Other scientists used dogs – particularly their slightly comedic scratching reflexes – as an avenue to explore involuntary reflexes and how the body's central nervous system processes and actions them. In fact, the observation of the dog's classic flank-scratch (which you can elicit by scratching a dog on the flanks below the hips) is a key part of an important turn-of-the-century text by Charles Scott Sherrington (1857–1952) called *The Integrative Action of the Nervous System*. In the book (which is still in print), Sherrington successfully argued that the body's reflex actions (think: hammer on knee) should be considered as integrated activities of nervous systems – essentially that a bundle of nerves is 'adapted to the needs of the organism'. He introduced the notion of 'reciprocal innervation': that, when one set of muscles is stimulated, the muscles that oppose this become simultaneously inhibited. He also considered how neurons communicate with one another, employing the term 'synapses' for the first time.*

Like Darwin, Sherrington used the notion of reflex action in dogs being scratched as a way to pull readers into his work. Also like Darwin, on the whole, he was loathe to cause suffering to the dogs in his charge. His former lab-boy would later recall: 'I was there for six years and during that time he used one dog; he wanted to do some cooling experiments, and when he finished he gave the dog to someone as a pet.' But Darwin and Sherrington were outliers. Not all laboratory dogs had it quite so lucky. Increasingly, to allay the cruelty being forced upon them, laboratory dogs needed human allies. In two dynamic Swedish women, they would find them.

* Sherrington also had a way with words: 'The brain is waking and with it the mind is returning. It is as if the Milky Way entered upon some cosmic dance. Swiftly the head mass becomes an enchanted loom where millions of flashing shuttles weave a dissolving pattern, always a meaningful pattern though never an abiding one; a shifting harmony of subpatterns.'

By 1900, five years after his death, Louis Pasteur had garnered something close to superstar status in scientific circles. People all over the world talked of his greatness. Like some sort of alchemist, he and his laboratory team had produced more than just cures for rabies or vaccines for anthrax. His laboratory set the stage for a host of marked scientific discoveries. Not only did Pasteur find a smoking gun in the cause of many diseases by pointing the finger at unseeable microscopic germs, but he also helped found the scientific pursuit of bacteriology, which would go on to save many millions of lives in the century that followed. He discovered the process of treating milk (and wine) to limit bacterial contamination: pasteurisation, the process that takes his name.

Following on from his death in 1895, Pasteur's laboratory staff committed to continuing his legacy and building on its already notable reputation in the international science community. In fact, so esteemed had Pasteur's laboratory become that it was often visited by dignitaries or members of the aristocracy, eager to learn more about the great scientific strides being made behind their closed doors, and how they might be applied to society to improve healthcare and save lives. One of the guests it received during this period was a confident, well-mannered lady with links to Swedish royalty whose name was Emilie Augusta Louise 'Lizzy' Lind-af-Hageby. In due course, the Establishment would argue that Lind-af-Hageby was a hysterical, rabid feminist. But there were other reasons to rather like her too, as you will see in the rest of this chapter.

Chief among Lind-af-Hageby's attributes were an amenable and friendly manner, a hawk-like attention to detail and an ability to keep all who conversed with her on their toes. When it comes to her character, you only need to know this: at one point, in a defamation case, she chose to represent herself. During the legal proceedings, she asked 20,000 questions of a cast of thirty-four witnesses over a period of nine and a half hours and spoke 330,000 words. This at a time when a woman could not be admitted to act as a lawyer because, under the 1843 Solicitors Act, women were not considered 'people' under UK law. ('She is a woman of marvellous power,' the judge later reported. 'Day after day she showed no sign of fatigue and did not lose her temper.') For this alone, Lizzy Lind-af-Hageby deserves some sort of remembrance statue. She didn't get one, however. That honour would go to the unnamed brown dog whom she would introduce to the world.

Like others of her time, Lind-af-Hageby's passion for dogs, and her desperate belief in their capacity to suffer, was partly inspired from her reading Darwin's *On the Origin of Species*. She saw in this book that the anthropocentric notion of man, ruler of all he surveys, was fallible. According to her notes, Darwin's ideas 'taught that if there is this kinship physically between all living creatures, surely a responsibility rests upon us to see that these creatures, who have nerves as we have, who are made of the same flesh and blood as we are, who have minds differing from ours not in kind but in degree, should be protected, as far as in our power lies, from ill-treatment, cruelty and abuse of every kind.'

Very little is known about the scale of animal cruelty that Lind-af-Hageby saw on her visit to Pasteur's laboratory (only that 'they had seen hundreds of animals dying in agony'*) but it was to have a lasting effect on her, so much so that she is alleged to have gone straight back to Sweden and joined the

* We have come a long way since the early 1900s. While animal experimentation remains a contentious issue, there is no denying the role it continues to play in medical research, under far stricter laws that see the use of anaesthesia and pain relief and appropriate care administered under the watchful eyes of ethics boards. Today, the Institut Pasteur has comprehensive information on the use of animals in its research, available on its website: www.pasteur.fr/en.

Nordic Anti-Vivisection Society. In fact, within a year, Lind-af-Hageby was its honourable chair. It was here, in Sweden, that she and her friend Leisa Katherine Schartau devised a stealthy and very cunning plan: they would infiltrate the laboratories of a well-known British medical institution and see for themselves the suffering of animals taking place there, all the while recording their observations in a secret diary that they would later publish.

The anti-vivisectionist movement faced a peculiar challenge at the turn of the twentieth century. As animals and their welfare continued to become an issue of public discussion, and dogs increasingly became animals of homes rather than streets, many laboratories took a simple approach: lest there be a scandal, they shut their doors and locked them to the interested public. Behind these closed doors they carried on their experiments on dogs, it was alleged, safe from public scrutiny.

Getting in would be tricky. Lind-af-Hageby and Schartau had to find a way in to an academic world that, being women, was largely closed off to them. But they spotted a loophole and, in 1902, they enrolled and became students at the London School of Medicine for Women, a college that did not practise vivisection per se, but that gave its students visiting rights to other London institutions that did. When opportunities arose – as they often did – the pair attended, notebooks in hand. In all, during their time in academia, the two ladies attended a hundred lectures and demonstrations at King's College and University College London. Of these, fifty involved experiments on live animals and many of them involved dogs. In all cases, the pair kept copious notes. Then, in 1903, their extracts were published. Their diary, when it was finally put together, was first called *Eye Witness* and later renamed *The Shambles of Science: Extracts from the Diary of Two Students of Physiology.** When published,

* In this context, the word 'shambles' is used to mean 'slaughterhouse'.

Shambles of Science caused a sensation – not least because it depicted scientists as entitled, cruel and totally detached from their animal subjects. One chapter was especially damning, featuring an unnamed brown dog being experimented on alive, without anaesthesia, while medical students laughed and jeered. The chapter was simply titled 'Fun'.

Needless to say, the tell-all account became something of a literary phenomenon. It was an instant hit, racking up more than 200 reviews in the British papers in the months that followed. But that wasn't the end of it; Lind-af-Hageby and Schartau went further. They organised rallies, they debated publicly with physiologists and doctors, they did interviews with the press, they spoke in the streets about what they considered to be a grave injustice. Some called it the first mass protest in the history of the anti-vivisection movement. Importantly, it was tuning in not only to national distress over animal cruelty but also to the feminist movement and more general frustrations felt by women at the time. 'The book *Shambles of Science* was printed [in] 1903, in a time when women did not have the right to vote, were not allowed to study to become lawyers, and when prominent medical scientists insisted that a woman who educated herself took the risk of damaging her uterus (and so could not have children),' writes Swedish author Lisa Gålmark.

In some ways, it was inevitable that such an account would find its way into the public arena and cause so much drama. After all, in 1878 there were 300 experiments known to have been performed on animals in the UK; by the time *Shambles of Science* was released, this number was exponentially higher – some say closer to 20,000 operations each year. Because of numbers like these, a succession of celebrities had been drawn to the cause, including Charles Dickens, Rudyard Kipling and Thomas Hardy, each adding further oxygen to the cultural flames.* Medical institutions rallied against their frustrations, pointing out their genuine good intentions and (what were then seen as) the high standards of their practitioners – that many, if not nearly all, of these scientists were in it to improve the prospects of those suffering great pain or injury.

In part because Queen Victoria had herself expressed disgust at vivisection, and in part because members of both houses of Parliament had their own reservations, the 1876 Cruelty to Animals Act had been passed to reduce tensions. The problem with this legislation was, its critics argued, that it had been watered down and all sorts of loopholes had been left open for institutions to exploit. For instance, under the legislation, animals had to be anaesthetised while being experimented upon, unless the animal's being anaesthetised might sully the results of the experiment, in which case it was permitted. This ambiguity could easily be seized upon by those looking to reduce laboratories' costs – anaesthetics were not only expensive but they required skill to administer, after all. Ambiguity existed in other parts of the law. Animals were not to be experimented on more than once, but multiple procedures were permitted on the same animal if it could be argued that this was part of a single operation. Another potential loophole that campaigners expressed concern about was the decree that animals were to be killed when a given study ceased, unless doing so would hinder the results of the given experiment. This led to allegations that some animals were being kept alive unfairly, under the blurred lines of when one experiment ends and a new one begins. Although, in name, the 1876 Cruelty to Animals Act was strong, its critics

* Hardy is said to have left a copy on his table for visitors to dip into while they waited, so that 'everybody who comes into this room, where it lies on my table, dips into it, etc, and, I hope, profits something.'

argued that the murkier elements of the scientific establishment had plenty of ways to get around it to keep their costs low and maintain a higher rate of discovery than rival institutions. And, because the doors of their institutions were often firmly locked to outsiders, no one could see – let alone challenge – any illegal activities anyway. For a long time, any institution that practised illegal vivisection was unlikely to ever be found out. *The Shambles of Science* went some way to changing that.

The chapter named 'Fun' was especially important in this respect. It was in this section that the titular brown dog of this chapter gained its notoriety as a *cause célèbre* for the anti-vivisection movement. In it Lind-af-Hageby and Schartau described a procedure they had observed performed upon a brown mongrel, whereby the dog's neck was opened to expose the salivary glands *à la* Pavlov. The aim of the procedure was to show students that salivary pressure operated independently from blood pressure. Readers of a more sensitive disposition may prefer to skip the next three paragraphs.

'Today's lecture will include a repetition of a demonstration which failed last time,' their diary entry read. 'A large dog, stretched on its back on an operation board, is carried into the lecture-room by the demonstrator and the laboratory attendant. Its legs are fixed to the board, its head is firmly held in the usual manner, and it is tightly muzzled.'

The key sentence here was the first: that this was a *repetition* of a previously failed demonstration, being performed on the same dog. Technically, Lind-af-Hageby and Schartau argued, this should have been illegal. But there was more.

'There is a large incision in the side of the neck, exposing the gland,' they wrote. 'The animal exhibits all signs of intense suffering; in his struggles, he again and again lifts his body from the board, and makes powerful attempts to get free.'

The animal clearly had not been anaesthetised correctly, they asserted. Additionally, their words alleged, the dog had been killed by an unlicensed research student after its procedure. There were plainly multiple failings and the law had not been correctly followed. This enraged readers of a certain demographic and tensions rose.

Particularly incensed was Stephen Coleridge (1854–1936), secretary of the National Anti-Vivisection Society. In a packed St James' Hall in Piccadilly, with more than 2,000 people in attendance, Coleridge pulled no punches. He accused the perpetrators – Ernest Starling, Professor of Physiology at University College London, and his brother-in-law William Bayliss – of nothing less than torture.* Hearing of the event, Bayliss demanded a public apology. When no apology was forthcoming, Bayliss sued Coleridge for libel and, within six months of the publication of *Shambles of Science* in 1903, the two would face off in a courtroom. In this showbiz trial, Bayliss, on behalf of the University College London, would win, the jury voting unanimously in favour of his exemplary medical credentials and his apparently good intention. According to the judge, Bayliss' name had been unfairly besmirched by what the judge called Lind-af-Hageby and Schartau's 'hysterical' accounts, an assertion many newspapers of the time refuted.

Coleridge was ordered by the court to pay £5,000 to Bayliss, which he duly obliged, but – thanks to a plea from *The Daily News* – within four months the public donated £5,700 to personally cover Coleridge's losses. And so, the fight continued.

Opinions were polarising: on one side was science, cruising into the twentieth century on a wave of enquiry; on the other was a different wave – empathy, compassion for nature, and

* 'If this is not torture, let Mr. Bayliss and his friends … tell us in Heaven's name what torture is.'

compassion, above all else, for dogs. A vocal minority of vivisectionists argued that the campaigners were displaying what we might now call 'virtue signals' to others of their social class. Others pointed out the hypocrisies of anti-vivisection campaigners – that many ate meat or wore garments made of fur, for instance. Some expressed frustration that vivisection represented a tiny fraction of a percentage of the animal suffering that occurred at the hands of humanity. Debate raged, but there was no place for them to play out, no physical manifestation of the problem. For that, there needed to be a symbol – a statue. If *Shambles* was the spark, this now-forgotten statue of an unnamed brown dog would be the lightning rod around which the frustrations of a century would gather.

As statues go, it would have been hard not to notice. The bronze statue of the unnamed dog stood atop a pillar more than seven feet high. It consisted of two fountains: one serving humans and horses with drinking water, and the other, lower one for passing dogs. Commissioned for £120 by the World League Against Vivisection, at first the statue had nowhere to go. No one would dare to give it a home. Eventually the London Borough of Battersea stepped up. According to the feminist writer Coral Lansbury, it was fitting that the statue's home would be Battersea. After all, the borough was already a broiling hotpot of radical ideas, where proletarians, socialists and other wannabe modernisers brushed up against one another, all agitated by poor working conditions, pollution, inadequate housing, fear of crime, a mistrust and resentment of high society down the road. As an indicator of Battersea's progressive politics, it was home to London's only anti-vivisection hospital, which performed all its duties without using animal experiments and aimed to employ physicians who supported the cause. It was also home, of course, to Britain's very first 'Home for Dogs'.

The list of attendees and speakers at the statue's unveiling says it all: among the line-up were George Bernard Shaw (playwright, activist, polemicist), Charlotte Despard (suffragist, socialist, pacifist), and the noted anti-vivisectionist Reverend Charles Noel, who led the crowd in an impromptu anti-vivisectionist hymn – an anthem named 'Ring the Bells of Mercy'. The singing over, the statue was unveiled and the frustrated crowds got to see the weatherproof manifestation of the brown dog with no name for the first time. Its inscription read:

In Memory of the Brown Terrier
Dog Done to Death in the Laboratories
of University College in February
1903 after having endured Vivisection
extending over more than Two Months
and having been handed over from
one Vivisector to Another
Till Death came to his Release.

Also in Memory of the 232 dogs
Vivisected at the same place during the year 1902.

Men and Women of England
how long shall these Things be?

The words were stinging. In fact, this was as close to baiting as polite society allowed. Clearly a line had been crossed. Understandably perhaps, medical institutions at the time were incensed but, crucially, so were other bastions of the Establishment. *The New York Times* referred to the statue as 'a slander on the whole medical profession' and said the inscriptions were the 'hysterical language customary of anti-vivisectionists'. Rioting became wholly foreseeable. Notably, it was the students of science, not the practitioners, who were most enraged.

First came the undergraduates, armed with crowbars and sledgehammers, who managed to dent the statue with a hammer before being duly arrested. Next came a larger band, made up of medical students from a range of London medical institutions. They walked London's Strand armed with miniature makeshift brown dogs on sticks and an effigy of the magistrate responsible for fining the first group of protesters with crowbars and sledgehammers.* Some medical students took to invading women's suffragette meetings, even though the interests of the two movements did not always overlap. (Not all anti-vivisectionists were interested in the rights of women and, vice versa, not all suffragettes cared for the rights of dogs.) In one fateful meeting, the medical students let off fireworks to drown out speakers and smashed up tables and chairs before being led out by a man, inexplicably, playing the bagpipes.†

On 10 December 1907 came another riotous rabble. This demonstration was supposed to be far better organised – thousands were expected. The plan was that an army of students would march to Battersea, topple the statue and throw it in the Thames, before proudly walking back to Trafalgar Square. Merchandise was prepared for this grand event. Street vendors would sell novelty handkerchiefs embroidered with the words 'Brown Dog's inscription is a lie, and the statuette an insult to the London University.' This

* They intended to light this effigy but, as often happens when protesters with specialist interests are forced to do novel things such as successfully light an effigy in a busy street, the flames failed to get going. They threw the effigy into the Thames instead.

† 'Medical Students' Gallant Fight with Women' was how the *Daily Express* reported the incident.

mass rally would put an end to the brown dog affair for good, the students hoped. Things didn't go quite as planned.

First, a large number of the protesters failed to show up. Instead of thousands, in the end just a hundred students made the march to Battersea. Worse, when they got there, they faced a mob of locals who fought back. The students' attempt to get near the statue foiled, they headed to Battersea's Anti-Vivisection Hospital to protest there instead. This was not a good move for an angry throng. During one particularly violent scuffle with locals, one of the students fell from the top of a tram and needed urgent medical attention. The nearest location for treatment was the institution he had just been shouting angrily at. Suffice to say, some there likely had on their face a wry smile.

In the months that followed, demonstrations and riots between the two groups continued. So much so that the Brown Dog statue was rarely without an armed guard. In fact, in some cases, the statue was attended by six police officers at a time, costing the constabulary a considerable £700 a year. The police department asked Battersea Council to foot the bill. They refused. The anti-vivisectionists thought that the vivisection laboratories should foot the bill. They refused. The British Medical Association argued that the statue should be removed and taken to the Dog's Home whereupon it should be hammered to pieces in front of anti-vivisection campaigners. ('If their feelings were too much for them, doubtless an anaesthetic could be administered,' they added, acidly.)

Within months, the writing was on the wall for the prized bronze statue. Alas, the cost of protecting the statue was too high. Rallies, letters, demonstrations, petitions – nothing could save the statue from its fate. Just before dawn on 10 March 1910, four council workmen – under the armed guard of 120 police officers – quietly removed the statue of the little brown dog. It was taken, rumour has it, to a local blacksmith and then melted down. This was a culture war of a different time. A different place. The same kind of magnitude.*

The Brown Dog Affair was something of a flashpoint in time. For a few moments it shone a hard light on society and revealed everything: the empowered, the angered, the frustrated, the liberated. The chasmic divide between science and society was there for all to see. One in which dogs were subjects. Instruments to some, instrumental to others. Subjects of science; subjected to it in the worst possible ways. In a whoosh of momentum, in only a matter of years Lind-af-Hageby and Schartau changed science's relationship with dogs, expanding the debate and chipping at the weak spots of societal divide throughout the continent. Many popular accounts of the history of science prefer to leave out the detractors of science and consider only those geniuses whose eyes are focused on driving forward our understanding of the world. I have included Lind-af-Hageby and Schartau's contribution in this book to show that, through energising public debate on an issue, they had just as much power to shape and influence the discoveries of future decades. To influence what we know of ourselves and other animals. To further thin the divide between humans and dogs for ever.

As Pavlov's ideas moved across continents, throughout Europe and into America, the debate that Lind-af-Hageby and Schartau ignited between science and society continued in Britain, and the contentiousness and apparent barbarism of animal experimentation would continue to be a key cause for the middle classes to rally against, at least until the Great War, when the public's moods and frustrations changed their

* In 1985, a new statue of the brown dog was erected in Battersea Park, funded by the National Anti-Vivisection Society and the British Union for the Abolition of Vivisection. It was later taken down, only to be reinstalled (now in its third incarnation) in 1994.

focus.* Partly for this reason (and partly because so many British scientists – retrained and flung into the trenches – had been killed), when dogs opened up new avenues of scientific study it would be on foreign shores that discoveries would occur – mostly notably in North America.

In North America, Pavlov's ideas would travel far and wide, mutating across college campuses and lecture theatres as they went. While many of these new strains would fall in the cracks and fail to find light, some seeds found fertile soil in the minds of a new generation of scientists who had lived through the horrors of war and were eager not only to understand how minds work but also to use this knowledge to make lives better. It was in these research institutions that the new sciences of neuropsychology, cognitive psychology and Behaviourism would grow. And they would grow richer still, for good or bad, in the presence of dogs.

---

* As renowned animal rights advocate Rod Preece puts it: 'Victorian and Edwardian optimism that solutions to ethical problems were just around the corner was dealt a vicious blow by the miseries of the Great War and the loss of confidence that ensued. Those who continued in their seemingly utopian paths were usually deemed eccentric, if not worse.'

SECTION II

FETCH, RETRIEVE

CHAPTER 5

Alpha, beta, doubter

> 'There is no such thing as a new idea. It is impossible. We simply take a lot of old ideas and put them into a sort of mental kaleidoscope. We give them a turn and they make new and curious combinations. We keep on turning and making new combinations indefinitely; but they are the same old pieces of colored glass that have been in use through all the ages.'
>
> – Mark Twain, *Mark Twain's Own Autobiography*

Looking across the length and breadth of dog breeds in the modern day, it is easy to be overcome by the varieties, the shapes, the sizes, the diversity of dogs. It is as if God has outsourced a small part of creation to the Jim Henson

Company. And, just as every Muppet is merely an exaggeration of some extinct nameless sock-puppet, everywhere one looks with dogs, there is the same story: an exaggeration upon exaggeration of an elusive extinct dog, a shared ancestor, a village dog who passed silently and namelessly from history.

The breeds of dogs really are staggering. Lesser-known examples of modern-day breeds include the Norwegian Lundehund, a perky urchin-like sprog bred to catch puffins; the Sloughi, a lithe sprinter able to bring down gazelle and hare; the Bergamasco Sheepdog, a preposterous nimbus-like cloud of a dog with thick strands of matted hair to keep it warm in the Italian Alps; the Finnish Lapphund, a reindeer herder; the Norwegian Buhund, apparent friend to the Vikings; the Plott hound, bred (in part) to spook bears from the ground and up into trees.

Dogs are incredible. Biologically implausible, somehow. Nothing else on Earth has so wide a range of characteristics within the same species category. There are tens of thousands of separate species of insect who so closely match one another that scientists have to spend many hours at microscopes investigating the insects' genitals to get a true taxonomic identification. Even then, these scientists sometimes get it wrong. This is absolutely not the case with dogs. A dog is a dog is a dog, regardless of breed.

Yet more incredible still is that all of this breed diversity in dogs has taken fewer than 5,000 years to come about, through selective breeding by humans, who consciously or unconsciously were acting in the same way that Darwin imagined, encouraging the breeding of desired traits, rewarding with reproductive opportunities the dogs best at hunting, chasing, pointing, guarding, fighting, digging – in other words, artificially selecting the dogs best up to the job.

The vast majority of breed diversity in dogs has happened incredibly recently. In the last 200 years, for instance, the number of dog breeds has seen a dramatic increase, from just fifteen or so known breeds in Britain at the start of the nineteenth century to almost sixty at the end. Today, there are estimated to be more than 400 breeds across the world. This sharp rise in breeds, particularly at the end of the nineteenth century, saw numerous 'kennel clubs' brought into existence to celebrate and promote these differences, and shows (of which the most famous is Crufts) to feed the growing appetite of a mass market. As interest in dogs grew in Britain and across the rest of Europe, so too did the demand for new breeds that made their owners stand out from the crowd. Many breeds became 'must-have' and the talk of the town – often the quirkier the better. *Punch* lampooned this strange trend for new breeds in an 1889 cartoon sketch that featured a well-to-do lady walking a range of perverted exaggerations of dog-kind – a dorgupine, a crocodachshund, a hedge-dog, a hippopotamian bulldog with giant tusks and so on.

Just as Dickens had previously caused the stock of dogs to shoot up across society, the twentieth century saw a certain 'Hollywood effect' begin to play out in the USA. Of this new crop of canine movie stars, most notable was the German shepherd Rin Tin Tin (1918–32), star of twenty-seven Hollywood films, who single-handedly (if that is the right expression) raked in the cash that caused the early domination of Warner Brothers over its rival studios. Then there was the German shepherd known as Strongheart (real name Etzel von Oeringen, 1917–29), who was something of a Tom Cruise of dogs and starred in films such as *The Silent Call* (1921), *Brawn of the North* (1922), *The Love Master* (1924) and *White Fang* (1925). Because of star names like these, new dog registrations went from just 5,000 in 1900 to 80,000 three decades later. At one point, according to anthrozoologist Hal Herzog, new dog registrations were growing at fifteen times the rate of the human population.

By the 1940s, ideas about what dog breeds did and did not represent were so established that marketing executives were using dog breeds in all sorts of ways to promote their products. Inevitably there began a period of cultural type-casting of dogs that continues to this day. Breeds weren't meaningless; they needed careful consideration. Many came with a newfound public image – a certain kind of status or reputation that could be exploited for profit by Madison Avenue. Resplendent German shepherds featured in adverts for Swiss watches and Buicks, for instance; rough and ready huskies for motor oil and gin; golden retrievers and terriers for family items like matches and beer, Cellophane, Ford, and menswear from Hart Schaffner Marx.* During this period, *Vogue* magazine developed a fascination for lap-dog breeds such as pugs and Pekinese, breeds that represented 'the idea of feminine sensuousness, luxuriousness and stylish livings', according to Annamari Vänskä, Professor of Fashion Research at the Aalto University, Finland. 'Love fashion, love her dog,' implored the title of one *Vogue* article.

It is clear from representations like these that dog breeds were each finding their own climate, their own culture, their own new niche within modern Western society. Predictably, a swirling mass of new financial interests sprouted off the back of the booming trade. From humble beginnings, entire pet industries were born. Dog-food manufacturers took off, as did veterinary services, insurance, dog toys, dog-friendly yards, dog-friendly parks. Among all this arose another industry – one that emerged like an apparition from out of

* These popular breeds would soon be followed by a newcomer – the poodle. From 1946 onwards, for a twenty-year period, poodle registrations went through the roof – in the USA, a 12,000 per cent increase in registrations for poodles with the American Kennel Club took place, spiking in 1969. According to Herzog, the AKC had to hire new staff to assist with the sudden increase in paperwork. Off the back of this, a trend for all things poodle arrived; along came poodle skirts, poodle soft drinks, poodle cheesecakes, poodle hosiery, poodle razors, cigarettes, souvenirs …

nowhere in the 1940s but that caught on quick: to cater for the needs of these dogs, the public needed to know how dogs work – how they think, how to get them to do what is expected of them, how best to train them. Hollywood's dogs provided some answers – the feats that silver-screen dogs managed at the hands of trainers were nothing short of staggering, after all. But science also began looking to the wild for answers on the question of training. If dogs were wolves made tame, they deduced, then might the secret of how best to train them be found in the blood-soaked tapestry of their wolfish past?

For decades, Pavlov's ideas about how behaviours manifest themselves had influenced popular thinking about dogs. Now, ethological ideas would be added to the mix.

Alphas were on the up.

'A bitch and a dog as top animals carry through their rank order and as single individuals of the society, they form a pair. Between them there is no question of status and argument concerning rank, even though small fictions of another type (jealousy) are not uncommon. By incessant control and repression of all types of competition (within the same sex), both of these "α animals" [alpha animals] defend their social position.'

These words come from one of the first and most widely cited case studies of wolves ever undertaken, published in 1947 by Rudolph Schenkel, animal behaviourist at the University of Basel, Switzerland, after countless hours of behavioural observations. Schenkel's paper 'Expression Studies on Wolves' became the cornerstone for anyone studying wolves (and, by extension, dogs) in the decades that followed. Unfortunately, once laid, cornerstones can be difficult to remove.

In his paper, Schenkel talked of the 'violent rivalries' of wolves seeking to be 'first in the pack'; the 'lead wolf' seeking dominance above all things; of the desperate competition for 'alpha' status. For a research paper, this was gripping stuff. It was bloody. It was fierce. The paper showed how wolves may have evolved and prospered through a life under constant threat of violence. The article is littered with keywords you will know – notions of alphas, control, repression. We rather take terms like these for granted now, but to mid-twentieth-century audiences these were exciting concepts in biology. Schenkel's ground-breaking account would be the document that connected readers all around the world with the lives of wolves and the behaviours that maintain wolf society. It was this information we would apply to dogs, their pliable descendants. Just as with Schenkel's wolves, we would see dogs as subordinate animals desperate to take the spot of alpha in our homes. We would come to accept that, in order to keep dogs in their place, we must take on the role of alpha wolf in the family – applying our will through brute strength, aggression, bullying, and often through force. The problem was that we were wrong.

The legacy of Schenkel's research paper plays out in lots of ways across society in the modern age, not least in its application to males – in the context of their access to power, money and apparently swooning mates; everything the weaker, submissive male (the beta male) cannot access. We only need look at politics and how the language of 'alpha male dominance' infects public commentary ('"Alpha male" handshakes as Trump, Kim meet' was the Reuters headline when the two world leaders met in June 2018); it manifests itself in certain workplaces ('Alpha male culture helped cause financial crash in UK, MPs told' the *Guardian* reported in February 2018), in sport (one of the leading training academies in the Ultimate Fighting Championships, founded by Urijah Faber, goes by the name of 'Team Alpha Male'), in royal commentary ('"Alpha male" Prince Philip was "never going to be easy option" as Queen's husband,' wrote the *Express* in November 2020), in fitness (just google 'alpha male' and scan through the images without slipping on the body wax) and in the normal day-to-day lingo of males of a certain demographic, such as when the UK's professional agitator Nigel Farage dismissed Trump's infamous comment to 'grab 'em by the pussy' as 'alpha male boasting'. The fact that googling the term 'alpha male' provides 12 million search results whereas 'alpha female' provides 1.5 million results tells its own story.

In the context of dog training, Schenkel's initial commentary on the dominance behaviour of wolves (and the generation of wolf scientists it would go on to influence)* had one notable legacy that still plays out today: that of 'alpha dog theory', where dogs are kept firmly 'in their place' through a range of practical, firm measures. Among the most famous proponent of this method of training is Cesar Millan, television's 'Dog Whisperer', a highly successful author and speaker in his own right to whom millions subscribe.

Key to managing the negative behaviours of dogs, experts like Millan argue, is to show the dog who's boss. You mustn't let your dog go in a doorway before you, Millan's subscribers argue. You must eat first, not the dog. Likewise, the dog should never walk before you on a leash. Choke chains, prong collars and shock-collars (collars able to give mild electric shocks on demand by remote control) are just some of the tools that supporters of this training method sometimes

---

* This is a nod to Dave Mech, the American biologist specialising in the study of wolves, who popularised many of Schenkel's ideas to the wider public. In 2008, Mech publicly denounced the term 'alpha' (to 'once and for all end the outmoded view of the wolf pack as an aggressive assortment of wolves consistently competing with each other to take over the pack'). He predicted that it would take twenty years for wider culture to catch up. We wait.

employ. The physical techniques some wield to get results include 'flooding' (forcibly submerging a dog's head) and 'alpha rolling' (pinning a dog down with force). With language like this, it is clear why such methods have proved so divisive in the dog-owning community.

Those who support this method of training argue that they have the best interests of their dogs at heart: by using a shock-collar, for instance, they can avert behaviours (such as killing sheep) that might see their dogs shot by farmers. But there is a deep flaw in this philosophy, which is that the ideas around 'alpha males' in wolf packs are somewhat bunkum. Misjudged. That is, according to many modern-day scientists. Alphas, as Schenkel viewed them, do not exist.

The problem was, scientists would later realise, that the wolves Schenkel first observed were not behaving as wild wolves behave. His were captive wolves thrown in with one another, many unrelated to one another, kept in an enclosure at Switzerland's Zoo Basel. Schenkel's wolves were behaving as captive animals often do: unpredictably. No wonder they fought. They were placed in cages with unrelated family members, causing social disarray. In short, it was anything but the wolf's natural environment. When scientists realised this decades later, by studying wild populations of wolves, it was too late. The ideas of dominance theory – of alpha males, of top dogs – had washed into society and could not easily be erased.

'Observations of captive wolf packs have led not only to mistaken assessments of wolf behaviour,' writes John Bradshaw in *In Defence of Dogs*, 'but also to fundamental misunderstandings about the structure of wolf pack families themselves, misunderstandings that have warped the popular conception of dogs as well.' The truth, says Bradshaw, is that 'alpha' behaviours are not about dominance, exactly. They are an emergent property in wolves – a status that 'comes automatically with being a parent'. There is no such thing as an alpha male or alpha female, in other words. There is only mother, father, offspring, pack.

This isn't to say that dogs and wolf packs do not form dominance relationships with one another, of course. They regularly do this. In fact, dominance behaviours are observable in all sorts of vertebrates. Observations of free-living dogs and wolves show clearly that there is the formation of social hierarchies, so Schenkel wasn't totally wrong in his observations. The key point is that the rung of the ladder on which a given wolf might sit can be considered as much a measure of its life-stage (for instance, whether it is a youngster deferring to elders or vice versa) as it can a measure of some all-knowing alpha who spends every waking moment subduing the rise to power of others in the pack. Dominance isn't always asserted through violence or bloodshed, either. Tensions between individuals can be managed in far more subtle ways, which is apparent in the everyday behaviour of both wolves and dogs.

To get a feel for what Bradshaw and many other canine scientists argue, it helps to explore the ecology and social set-up of the grey wolves that today haunt the Northern Hemisphere's cooler regions. For in them, we can assess what wolves are and how their societies function, and then make some comparisons with dogs.

First, the basics. Hunting larger animals is a tough business for solitary species, so in some animals (such as killer whales, African hunting dogs and hyenas) it pays to work together to bring down prey. When these ecological situations arise – when, for instance, large prey becomes the only game in town – natural selection appears to favour the fortune of those individuals that work best in teams, and so traits for co-operative behaviour flourish and sociality becomes *de rigueur* within populations. Social hunting establishes itself within the behavioural repertoire of a species, not because animals become inherently good or slavishly empathic or somehow biologically 'woke' to one another, but simply because going it alone is akin to genetic suicide. The vast majority of those that try it may well fail to produce enough offspring to carry on the trend. This is the ecological niche in which wolves have found themselves.

Most wolves are part of packs. Packs protect wolves from starvation. But packs provide another, unexpected advantage: if pack-forming behaviour evolves across populations, a pack protects its members from attacks by other packs and provides territorial advantage. But packs are not easy to maintain. Each pack is filled with individuals, each with their own goal of maximising the transmission of their genes to future generations. It takes management to keep them together. It takes communication, much of which requires a top-down approach.

Most wolf packs consist of a breeding pair surrounded by younger family members who are not yet breeders themselves. In genetic terms, by fighting on behalf of mother and father and their fellow siblings, a successful non-breeder in the group helps the transmission of his or her genes into the next generation via the survival of close family – this is known as kin selection. Yet the strong bonds that exist within a pack do not last for ever, and eventually a non-breeder in the pack may seek a different strategy. They may choose to usurp mum or dad and have a go at raising their own cubs – creating their own pack, as it were. Going it alone brings with it many dangers. The first is the risk of starvation: finding food without an adequate pack is hard. The second – and by far the more serious – is that these animals are unable to defend a territory and, in the worst cases, they become lone wolves.

Lone wolves are treated with immense unease should they accidentally stray into another wolf's territory. In some parts of the United States, for instance, these territorial battles are the single biggest threat to wolves, with more than half of all wolf deaths occurring at the hands (jaws) of another wolf. For this reason, a wolf seeking to take control of its own destiny needs to make sure of certain things: first, plenty of food; second, an opportunity to mate; third, a territory to defend. Only with these elements available might a young wolf take its opportunity to splinter into a new group; even then, most will fail. With social animals, especially wolves, evolution rarely favours the brave. Instead, it favours those who understand others: those who understand how to back down so as to avoid injury when tensions run high, for instance, or those who have become experts of appeasement, or gentle when necessary and loyal when it matters. It favours experts in the habit of not rocking the boat too much or too often, lest they be ejected. It is this powder-keg environment, over many thousands of years of wolf evolution, that has given dogs many – but crucially not all – of the cognitive tools they now apply in our day-to-day interactions with them. It is also for this reason that dogs are such elegant masters of clear communication. In a society where chaos can be unleashed with one fiery glare, it's wise to get your messaging straight.

To keep the peace (or bide their time, depending on your perspective) wolves rely on a number of social cues that aim to keep the pack from erupting into all-out war. Many of these behaviours you can see in dogs. A breeding male wolf, for instance, will be greeted by all members of the pack with puppy-like gestures of submission such as face-licking (used by puppies to stimulate the mother to regurgitate food) and rolling over to expose the belly. Tails wag furiously in these moments. These behavioural adaptations signal to other members of the group a state of cool acquiesce – happy wolf vibes – that keep the boat from rocking too much. Interactions like these help groups maintain strong bonds, increasing their chances of making a successful kill. Yet still, each individual in the pack is a genetic island, although a bridge, of sorts, exists between them.

It is fair to say that a great many of the things that dogs do in their day-to-day behaviours were bestowed upon them during their history as wolves. Every dog breed alive today is something of a master of a few single notes in the wolf's rich orchestra of highly complex social behaviours. The Border Collie is a good case in point. As is well known, the breed has a fixation for sheep. Thus, when a Border Collie sees a sheep, it freezes and stares with a kind of laser-like intensity unapplied to any other object. This pointed glare is not an evolutionary invention to curry human favour, but a hyper-exaggerated predator stalk, used regularly by wolves when they hunt. Likewise, the familiar 'point' of some dogs, whereby the dog freezes and literally points its muzzle towards prey, is another part of the wolf's natural repertoire. Retrievers love to chase, as do sled-dogs, of course. They happily grasp prey too, yet very rarely do their behaviours progress to the killing part. Instead, retrievers are playing out little scenes from their wild history. They are playing (in the most literal sense) micro-routines of behaviours that we see in wild wolves today, like a kind of evolutionary *déjà vu*.

Other wolfish behaviours are evident in the day-to-day lives of dogs. The cock-legged stance of urination is one example, where small amounts of urine are squirted upon prominent objects within a territory, such as trees, rocks and fire hydrants. This behaviour is, in many ways, more than simply a wolf thing – in fact, very public acts of urination are a classic carnivoran strategy for social communication. Often special glands in the anus add extra notes to these chemical signals, communicating messages that may instil in others feelings we cannot imagine. These urine-marked spots are like status updates to any other wolves that may pass through, offering up exciting chemical messages akin to titillating Twitter updates.

Another of the vestigial wolfish behaviours that dogs engage regularly in is digging. Not only the behaviour of digging but also the *way* that dogs dig holes, with front legs to scrape and back legs to kick soil back away from the hole. This is a behaviour that well-fed dogs have no need for but that serves wolves well in the winter months when food may be scarce and some morsels (especially bones) may need burying and later revisiting.

Tail-wagging is a particularly wolfish behaviour. The dog's tail is, quite literally, a flag. In general terms, the more the flag waves, the more everyone else knows a dog is gleeful and content and in no way seeking to overthrow any breeding pairs and/or inadvertently start a turf war or any violent battle to the death. Tail-wagging is just one mammalian remedy that natural selection has favoured over time to ease social tensions. Some mammals use a different technique. Baboons, for instance, might seek to groom one another to diffuse tricky social situations in the group. Humans smile, chuckle or laugh. Bonobos, well ... let's just say they go in for a different kind of social lubrication. The result, for each species, is the same: tensions between group members subside and life continues safely for all parties.

The tails of dogs can also signal other emotional states. Tails held firmly backwards being waved tightly back and forth is a sign of gentle curiosity. A dog with tail held erect over the body suggests unpredictable states of anxiety and possible arousal. Speed of wagging also matters. Tiny, urgent wags occur when a dog is aware that tensions might be about to flare up. Big, gangly wags are the opposite – wags like these are quite simply the stuff of canine delight. Remarkably, research suggests that even the direction of wag communicates information to other dogs: if a wagging tail generally leans to the right, the dog is feeling friendly and happy, whereas a left-leaning wag is an indicator of nervousness and possibly stress.

Dogs show off a host of other methods and modes of communication acquired from their wolf ancestors. These include active submission (tail between legs), passive submission (lying on the back, belly exposed), submissive 'smiling' (curling of the corner of the mouth), tooth-baring threats (curling of the lip), and the play-bow (paws forward, head and shoulders lowered). These behaviours are little more than signalling devices. Their aim is to communicate to others their mental state, clearly and quickly. They can be employed for fun or in fear, to de-escalate tensions or to warn others they are close to the edge. The presence of these adaptations in both dogs and wolves today sheds light on the hard times these canids have endured evolutionarily.

With such a wide variety of behaviours shared across both dogs and wolves, it is easy to see why Schenkel and the biologists he inspired looked across at wolves and decided that their social systems were to be the blueprint for the dog's. It was, however, a mistake because, as well as the similarities, dogs and wolves clearly have many differences in the ways that they behave.

Among the most cited differences between dogs and wolves are those exposed in the now-famous experiment by the Swedish behavioural biologist Erik Zimen, detailed in his 1981 book *The Wolf: His Place in the Natural World*. Under identical conditions, Zimen and his colleagues raised a pack of poodles and a pack of wolves, cataloguing and logging the appearance of 362 behaviours (tail-wagging, stretching, play, howling, *etc.*) displayed by each pack. Of the behaviours exhibited by the wolves, only about 64 per cent were exhibited in the poodles. Compared to the wolves, the poodles were not able to take any larger prey and showed little motivation to do so. They chased, of course. But they chased all sorts of things – litter, leaves, birds. At best, their behaviours were undirected and uncoordinated. At worst, it was more like play. The poodles were far less likely to bare their teeth to defend themselves, too. They also chose to hang around with one another while sleeping – unlike wolves, who preferred to sleep alone. Overall, for at least one third of the time, poodles would lie together. Even when it was hot, the shared sleeping arrangement continued.

The results of Ziman's experiment suggested poodles were easy-going in their nature in a way that the wolves were not. The wolves appeared to have the cognitive wherewithal to communicate, coordinate and kill. But if the poodles had these skills, they did a great job of hiding it. No, theirs was a blissful kind of ignorance, at least for much of the time. They did not form an obvious pack or even manage much in the way of coordination at any point. Left to their own devices, the poodles became much like the village dogs we see today all over the world. Though they were clearly capable of observing status, they were inhibited by a joyful lack of ambition. The author Stephen Budiansky describes this state eloquently, calling free-roaming dogs like these 'a happy band of lunatics'.*

Another difference between dogs and wolves, only discovered in 2019, is how they use their faces. Without putting too fine a point on it, dogs can give us the puppy-dog eyes. To pull off this move requires the use of a specialist muscle – the *levator anguli oculi medialis* muscle (LAOM). In wolves, the muscle is so small it is almost non-existent. In dogs, however, it is positively pumped. In fact, so ripped is this muscle that it can pull the eyelids up to expose the whites of the eyes (the sclera). The discovery of this amazing musculature came courtesy of Juliane Kaminski and colleagues at the University of Portsmouth in 2019. According to Kaminski, their findings 'suggest that expressive eyebrows in dogs may be a result of humans' unconscious preferences that influenced selection during domestication'. Look into

* 'They are like an insane asylum full of inmates, all of whom believe they are Napoleon,' Budiansky observes wryly in *The Truth About Dogs*. 'Every once in a while they ask the guards if they can get together and have a Napoleon convention.'

those puppy-dog eyes in just the right way and, in more ways than one, you will see yourself reflected back. Whether we meant it or not, our species bestowed upon dogs the magic of facial expression – or at least this one example of facial agility.

In the tens of thousands of years that our two species have shared, dogs have come to differ from their wolf compatriots in another notable way: dogs of most breeds go through a longer development phase than wolves, with their 'childhood' occupying a longer proportion of their life. By the time most wolf pups are engaged in biting, pulling and various other acts of rough and tumble, for instance, most dog pups are comparatively low-achieving.

In summary, there are clearly differences between dogs and wolves. The wolf's day-to-day survival depends on the wolves who share their ecological space. The dog's survival, to a greater or lesser extent, depends upon us. So how did Schenkel get it so wrong?

In the epigraph of this chapter, I quote Mark Twain in his assertion that there are no new ideas, only ideas from generations beforehand, reframed or rejigged or recombined with other ideas, like coloured glass in a rotating kaleidoscope of human culture. In this context, it is highly likely that Schenkel was drawing on the coloured glass that was drifting throughout popular culture in his own age – most notably, how dogs appear to respond to punishment and how their behaviour can be altered through it. Clearly dominance relationships were drifting around in the ether during these decades.

At around the same time, Schenkel's contemporary, the Nobel Prize-winning Austrian animal behaviourist Konrad Lorenz, released his popular book *King Solomon's Ring* (1949) in which he put forward his own assertions for the dominance relationships between animals of the same species. In the book, he detailed how aggressive and powerful actions performed by dominant dogs saw weaker dogs give ground or roll over in an apparent act of submission. He also noted how humans can manifest with their dog a similar dominance relationship – that, if under duress, dogs respond to an owner as they would a dominant dog.* Clearly, Lorenz knew the impact of enforcement, punishment and brutality in influencing the behaviour of animals – particularly since, after being sent to the Russian front in 1944, he ended up becoming a prisoner of war in a Soviet facility for four years.

But Lorenz and Schenkel didn't pluck their ideas out of nowhere. It is highly likely that both were influenced by other great thinkers and practitioners of the previous generation, most notably in neighbouring Germany where enormous strides in understanding dog behaviour had been made at the hands of Colonel Konrad Most, a gentleman many refer to as the 'father of modern dog training'. Konrad Most started off training police dogs in 1906 but, within a few years, his apparent flair for working with dogs was spotted by those in higher office – particularly in the military. By 1912, Most was director of Berlin's State and Training Establishment, where his role was to create a small army of 6,000 trained dogs that could be used in upcoming wars. Just two years later, his team had a chance to prove themselves at the start of the First World War, where his canine graduates were deployed for a variety of tasks including keeping sentry, searching for casualties among fallen buildings, finding explosives and ratting out the trenches. In all, the military made use of 28,000 dogs trained by Most and his team.

Most's training style was firm and, in modern terms, physical. It depended upon an array of tools that were somewhat the fashion of the day, including a leash that was made rigid at the end so that it could be employed as a whip

---

* Some have argued that certain themes of the book were veiled references to Lorenz's private sympathies for the Third Reich. Lorenz was not only a member of the Nazi party but also a member of its Office for Race Policy.

should a dog fail to behave as it should. Few could deny that, on the surface at least, his methods seemed to work but to modern eyes they were horrifying and tragically effective.

After the war, Most's achievements saw him promoted to director of the newly created Canine Research Department of the Army High Command, a role he held until 1937.*

In his training regimes, Most stumbled upon a number of concepts that were then being discovered by biologists, including ideas about reinforcement that almost directly mirrored Pavlov and the psychologists he inspired. In fact, if Most had been a scientist rather than a military man, he may well have become a household name in popular science. Indeed, Pavlov himself recognised this, writing in 1928: 'It is evident that many striking instances of animal training belong to the same category as some of our phenomena, and they have borne witness for a long time to a constant lawfulness in some of the psychical manifestation in animals. It is regretted that science has so long overlooked these facts.' Many dog trainers of the subsequent decades were directly inspired by Most's work – particularly the publishing of what many consider the first dog-training manual ever written: 1910's *Training Dogs*.

In the decade that followed this, a number of German dog trainers made a move to the United States, establishing their own training schools in the tradition of Most. One of them, Carl Spitz, made his name in Hollywood by training dogs in film, most notably by training Buck in *The Call of the Wild*, with Clark Gable, and Toto for *The Wizard of Oz*. He would

---

* By the time the Second World War had broken out, the Americans were investing heavily in their own Dogs for Defence (DFD) programme. In the years building up to the war, the DFD recruited 40,000 dogs, of which 8,000 were re-homed due to ill temperament or physical impairment and 10,000 were deployed for the war effort, while 3,000 made it back and were discharged to family homes.

go on to set up America's own war-dog programme, where he would be joined by another German, Bill Koehler (who would later go on to become Walt Disney's head animal trainer).

Like Most, Koehler took something of a tough approach. Among the techniques he advocated were 'helicoptering' dogs into submission (don't ask – it's awful), drowning and the use of choke chains. 'In training, Koehler advocated letting dogs make mistakes, providing consequences for those mistakes and then providing praise or reward for desired behaviours,' writes dog trainer Jay Gray in *The Academic Journal of Canine Science.* Within decades, more than 40,000 dogs would be trained under Koehler's tutorage. To a degree, his legacy is still playing out in the modern-day 'alpha dog' theory that Cesar Millan and others espouse.

In recent years, many vocal proponents of the dog cognition community have raised issue with the 'alpha dog' approach, and their voices rang loud and clear while I researched this book. 'One can agree to respond to controls imposed by someone of higher status, but this is done, not out of fear, but out of respect and in anticipation of the rewards that one can expect by doing so,' writes Stanley Coren, the well-known dog behaviourist and professor of psychology at the University of British Columbia. John Bradshaw – himself a well-respected expert in human–dog relations – subscribes to a similar philosophy: 'Discipline in the sense of control, not discipline in the sense of punishment, is what is needed.' Renowned dog trainer Victoria Stillwell argues that, 'if your dog feels good about you, she will be happier, confident, better behaved, and more inclined to respond to you when you ask her to do something.'

Not everyone is so tempered on the issue. The psychologist and animal welfare scientist Lauren Robinson calls the alpha theory 'outdated and not based on science'. She is one of many to inform me of this.

And this isn't just an American phenomenon. 'I think the alpha-dog training myth is just as much of a problem in the UK as anywhere else,' Naomi Harvey, pet behaviour specialist, tells me. 'Although awareness has grown among professional bodies, the way dogs are represented in the media and TV is still outdated and problematic.'

'It's hard to know how long it'll take to wash out of the dog training community,' says the carnivore biologist Gabi Fleury, 'but there's been a recent (and welcome) trend towards positive-reinforcement trainers over dominance training.'

Scientists like these favour the application of operant conditioning in different form: a more positive manner and from a more compassionate arena. They see Most turned on his head – training through rewards not whipping rods. But even still, these modern scientists are people. People with minds and experience, not immune to occasional influence by the fleeting reflections of coloured glass laid down in the kaleidoscope of the past. We are all capable of making the mistake that Schenkel made. We are all fallible. Knowing and accepting this is what makes great scientists even greater. I say this as someone who loves science – as someone who really believes in it. Like democracy, science is the best system we have. To get the best of it, to see the clarity and the detail and true beauty of life, one must keep asking questions and challenging the wisdoms of old. This principle is true of all of the arms of science. Here, through dogs, we see a handy example of the societal consequences of its misapplication.

CHAPTER 6

Skinner, unboxed

> 'To know when one does not know is best.
> To think one knows when one does not know is a dire disease.'
>
> – Lao Tzu, Tao Te Ching.

By the 1950s, the ideas of both Pavlov and Thorndike were lumped together and somewhat refined by a third character, the American psychologist Burrhus Frederic Skinner (1904–90) – commonly referred to as B. F. Skinner. Skinner was the founder of the radical behaviourist school of psychology, a subject on which he published more than 180 articles and more than twenty books across his career. He was a kind of academic hardliner who saw nearly all animal actions to be operantly conditioned, even human actions. In fact, so entrenched did Skinner become in this view that he ceased to believe in any form of free will at all. His ideas curried immense favour in psychological circles at the time, intent as they were on reducing big ideas to singular underlying principles. Like Conwy Morgan and his famous Canon, the idea that all animal behaviours could be broken down into the same simplistic chunks was an elegant proof of Skinner's genius. He wore this level of reductivism like a badge of honour.

It won't surprise you to learn that Ivan Pavlov was something of a role model for Skinner, who it is said first came across the Russian academic juggernaut while reading a review of Pavlov's book given by none other than H. G. Wells, the acclaimed science-fiction writer. Wells could see the value of Pavlov's work (he likened his experimental achievements to 'a star which lights the world, shining down a vista hitherto unexplored') and his glowing write-up was enough to change the direction of Skinner's interests for ever. Still a young man, Skinner gave up his academic interest in English literature and instead turned to psychology.* Two years later, he was at Harvard, sitting in a packed-out auditorium and watching in awe as Pavlov himself delivered a keynote speech. After the event, Skinner queued up for his autograph (which he later hung on his wall) and, in no uncertain terms, a baton of intellect was passed between generations that would influence our understanding of the minds of animals – particularly dogs – immeasurably.

Like Thorndike, Skinner was keen on the use of puzzle boxes to understand his research animals. The Skinner Box, as it was known, was a kind of shrunken experimental arena in

---

* Skinner's interest in literature would resurface in his sci-fi novel, *Walden Two*, in which he imagined a world devoid of free will where human societies could be engineered by their environments to generate something very close to a utopia.

which rewards (or, less commonly, punishments) could be dished out in all sorts of ways. Two classic Skinner box designs were a box with a lever for small animals (mostly rats) to press, and a box with a coloured disc on one wall for birds (pigeons, mostly) to peck. Food rewards were a key component of both.

This experimental design, which could be tinkered with in order to modify a number of variables, helped Skinner and his team explore the senses of their research animals in ways that had never before been trialled. It helped him record and understand more about what his animal subjects observed and did not observe, for instance, and how their experiences of the world influenced future routines. He could explore the impact of rewards or punishment on behaviours and routines, measuring their response rates and behaviours in the same way that Thorndike and Turner had earlier popularised – namely, graphically.

Among the things Skinner is most remembered for is the principle of reinforcement. Simply put, the principle of reinforcement says that, if the consequence of any action is good, the probability of that action being repeated is strengthened – a concept known as positive reinforcement. Likewise, if the consequences of an action are bad, the chances are that the action will be avoided in future – negative reinforcement.

Like dog trainers of the time (including William Koehler, who was by then becoming the 'go-to' animal trainer in Hollywood), Skinner understood that punishment worked as a negative reinforcer in animals, but he chose to take his research in a different direction. On a practical level, negative reinforcements led to behaviours that made his work more complicated – it produced escape or avoidant behaviours that he considered undesirable. But he also saw the importance in gathering data from positive interactions with his animals, mostly through the provision of food rewards. It got results, essentially. 'Dr Skinner's was a world without punishment,' wrote the science writer Dava Sobel. 'Dr Skinner helped to shape behavioral psychology as both a laboratory science and a cogent philosophy.'

Skinner had other legacies that have long since been forgotten. One of these was his observation of unusual animal superstitions. This – particularly in reference to dogs and how they came to be understood – is worthy of a brief digression.

In the course of his laboratory work, particularly when observing animal subjects in his boxes, Skinner occasionally noticed a strange phenomenon: that animals sometimes developed ridiculous behavioural tics derived and maintained through the process of reinforcement. In 1948, Skinner published an article on the subject, in which he detailed the strange behaviours of pigeons in his Skinner boxes. He noticed that some of his research pigeons would come to make repeated turns in their cage in the mistaken belief that doing so would provide a food reward. One pigeon, he observed, became somewhat addicted to swinging its head like a pendulum, apparently in a bid to acquire food that was set up automatically to be dispensed. Apparently, through these strange behaviours, the pigeons were trying to influence the feeding regime. It had worked once – accidentally – and so the behaviour had stuck. These behavioural fallacies were, Skinner decided, superstitious.

Interestingly, once established, Skinner found that superstitious behaviours were extremely challenging for his study animals to shake off. A pigeon, for instance, might try out the same nod or twirl some 10,000 times without the 'reinforcement' of reward, yet still it might pursue this strange routine, determined (apparently) to see it pay off. Inevitably, and probably quite correctly, Skinner deduced that many human superstitions occur through the same pattern of events, where happenstance attracts habit.

Animal superstitions play out across a wide range of species. A guinea pig may develop the mistaken notion, for example, that shaking their heads in a particular manner is what gets them fed. In horses it can lead to pawing, where the horse draws a hoof along the ground as it is being fed and comes to condition itself operantly to the behavioural tic.

Superstition is certainly common to dogs of all breeds too, though more often than not these superstitious behaviours take on a more subtle form. There's not the same robotic jerking or jarring tics of the head, for instance, but the connections between action and apparent effect are commonplace. Dogs clearly see patterns and change their behaviours accordingly, in all sorts of way. My childhood dog Biff (who had a troubling puppyhood, we suspect) was fairly passive most of the time, but if he was shocked by a passing bike he would lunge and tear on his lead, barking wildly with teeth bared at the terrified cyclist, much to my embarrassment. Yet Biff was playing out what, to all intents and purposes, was a winning strategy:

1. see human riding metal contraption
2. bare teeth and bark at human riding metal contraption
3. successfully avert attack by human riding metal contraption

First time it worked. Second time a charm. Third time, again. Biff's behaviour was, in Skinner's words, reinforcing itself each time.

It isn't hard to see why animals so quickly come to exhibit behavioural routines like these since, in a bizarre sort of way, they work. The reward that Biff had was seeing the bike receding into the distance, his sense of fear melting away, and his uninterrupted walk with me continuing.

Another example of this kind of behaviour is when dogs attack postal workers who stray too close to the door with our letters and packages. Each and every day, all over the world, those incredibly committed dogs are 100 per cent successful: the mysterious human with the big bag always retreats once the dog courageously barks or scratches at the door. In this way dogs become something like the apocryphal story of the British man who throws freshly cut flowers onto his front yard each day to scare away tigers. 'I've been doing this for years,' he says. 'It really does work!' Can't fault the logic, one could argue.

In modern times, neurologists have acquired a good understanding of the pathways that occur in the brain when operant conditioning of this kind occurs. Dopamine, a hormone and neurotransmitter produced in the brain, seems to play a key part. Most people consider dopamine to be a chemical that gives us a warm, fuzzy feeling of contentment, but it is also the tool that helps our brain perceive the desirability or averseness of a given situation minute by minute. In the words of pharmacologists, it confers 'motivational salience' upon us. Without this molecule, one loses an ability to navigate goals moment by moment. In this way, through hormones, our brains are physically changed by circumstance – the more we repeat behaviours associated with pleasure, for instance, the more dopamine channels are activated, ready for the next hit. It is this, perhaps more than anything, that leads to behaviours becoming reinforced.

Timing in this process is critical, of course. For every single encounter that an animal has with its environment, there is a specific window in which minds are open to link cause and effect and behave accordingly. These moments really are short.* Crucially, this physiology tallies up with the basic principles of Behaviourism: that experience shapes behaviour.

As well as being a prolific writer and a key founding father of the behaviourist movement, Skinner's legacies include some other gems that it would be remiss of me not to include here. The first is that, upon watching a classroom of children undertaking a maths test at the behest of a teacher, he decided to invent a machine that could do the teacher's job more effectively.

---

* This has important ramifications for the dogs in our care. It means that if your dog tears up your sofa while you leave the house, coming home hours later and telling him or her off does nothing positive. Your dog cannot join the dots between past cause and effect. The window has passed. To the dog, you're just an increasingly unpredictable person standing pointing at objects and shouting. This does not help matters. If anything, unpredictable outbursts like these are likely to stress the dog even more.

This drew on Skinner's discoveries about operant conditioning. His idea was that kids in a mixed class of differing abilities would improve much more effectively if they were given immediate feedback on their maths performance. To manage this, he designed a kind of maths-teaching machine that offered incremental feedback after each problem, paving the way for a learning process known as 'programmed instruction' where learners use books or equipment, testing themselves throughout, without teacher guidance. The technique is still drawn upon in some classrooms today – indeed, it has had some resurgence since the Covid-19 pandemic, where many schoolchildren had to become distance learners, much to the horror of their parents.

Another of Skinner's gems is a little more out-there: at the request of his wife, he invented – for want of a better term – a kind of experimental tiny terra-dome for babies. This 'baby tender' was a kind of heated crib within a plexiglass bubble, which offered a safe and practical solution to catering for the needs of the baby. So proud was Skinner of this creation, that his own daughter spent much of her early life there. The downside to this was that, when the world's media covered the story of Skinner's Baby Tender, they would conflate the 'Baby in a Box' (as one headline put it) with his own patented experimental set-up, the 'Skinner box' (the problem-solving chamber for rats and pigeons that led to many of his research discoveries). This created rumours that he was experimenting on his own children – an allegation that would chase him around for decades despite being totally false.*

* This gave rise to all sorts of scandalous gossip that spread through lecture theatres the world over. The most notable was that the child had later become psychologically unstable and taken her own life in a bowling alley – a rumour that was still doing the rounds in published biographies of Skinner as recently as 2004. 'My early childhood, it's true, was certainly unusual – but I was far from unloved,' Deborah Skinner Buzan told the *Guardian* at the time. 'Call it what you will, the "aircrib", "baby box", "heir conditioner" (not my father's term) was a wonderful alternative to the cage-like cot.'

Maths and maternity aside, Skinner had a further legacy. It is one that barely gets a look-in among his other achievements but it is one that has important relevance to dogs, specifically in the way they associate events with behaviours. As well as pigeons and rats, Skinner trained up two protégés – Marian 'Mouse' Kruse and Keller Breland – in his laboratory who would fall in love (after a meet-cute involving a rat biting a finger), marry and later pluck up the courage to leave academia, armed with Skinner's learnings, and start up their own business called Animal Behavior Enterprises (ABE). Together, these two scientists influenced greatly what we know of how dogs learn and how best they can be trained.

Skinner got to know Kruse and Breland during the Second World War, when Skinner's skill at working with animals was called upon by the US government who saw something of a unique application of his ideas – namely, Project Pelican, in which Skinner was asked by military personnel to develop pigeon-guided missile systems for pinpoint bombing. To satisfy this task, Skinner hired the pair, cutting short their PhDs. During this period, seeing how animals could be trained to do all sorts of incredible things, Kruse and Breland came up with the idea for their new business venture.

If you have ever seen vintage black-and-white footage of an animal in a comedy costume doing something vaguely ridiculous, it is likely to have been thanks to ABE. In the 1950s, if you needed a chicken walking a tightrope, for instance, ABE was your go-to company. The same goes for animals dancing to jukeboxes, rabbits kissing plastic girlfriends or riding fire trucks or spinning wheels of fortune, or ducks playing the drums. Gigs like this were, for a long time, ABE's bread and butter. Theme parks hired them to train their animals. There were events, museums, Hollywood films. They made *The Today Show, Tonight Show, Time Magazine, The Wall Street Journal.* One of their adverts for Coast Federal Savings, featuring a rabbit hoarding money in a bank, ran for an astonishing twenty years. At their peak, the Brelands (now married) were training more than 1,000 animals each month and reaping the benefits financially. In terms of size and scale, there was no organisation that came anywhere close to ABE at the time. And they had Skinner to thank for it.

Of course, in modern-day terms, many of the behavioural feats that the pair achieved under the guise of their enterprise would now be viewed as ethically problematic. Only the most blindly enthusiastic sports fan would argue, for instance, that a raccoon playing basketball is not in some small way demeaning for the raccoon (and the sport of basketball). But the Brelands really did show just how much animals could achieve with the right sort of training – through positive reinforcement rather than through punishment or, in other words, through carrot not stick.

Central to the ABE training regime was the idea that rewards had to be provided as soon as possible to the desired behaviours – almost instantly – for an association to be strengthened. This is easy for animals at close control, such as rabbits and rats, where a treat can be handed over upon the correct response. But it would prove harder for animals performing actions at a distance from their trainers, where treats are not immediately as forthcoming. The Brelands devised a new methodology to cater for these moments: a kind of remote-control reward system. To this end, the pair invented 'clicker training' – a technique that would immensely improve the ways in which dogs and other animals could be trained. The Brelands considered their clicker a 'bridging stimulus' – a secondary reinforcer to increase the probability that a desired action be performed again. The technique took a while to bed in (a few decades, truthfully) but then took off spectacularly.

The clicker itself is a small curved bit of metal that, when pressed, emits its sound. Crucially, the sound of the click is distinctive from other noises that a dog may hear in its day-to-day life. Clicker training works by a trainer pressing the clicker at the exact moment a dog does the correct behaviour, so as to strengthen the association between action and reward. In the PDSA's words: 'the sound of the click is used as a "mark" for good behaviour. It pinpoints the exact moment your dog has done the right thing so they don't get confused.'*

Today the use of clickers has become a vital tool for the dog-training community. Bridging stimuli (such as clickers and whistles) are used in guidance- and assistance-dog training, sheep-dog trials, dog agility, bomb-disposal dogs and, most recently, dogs being trained to sniff out coronavirus on hospital patients. Clickers are, quite literally, life-saving. We have Skinner and his former colleagues Marian 'Mouse' Kruse and Keller Breland to thank for them. Theirs was a crucial application of Skinner's ideas – applied science, for the masses.

Not all of Skinner's ideas were treated with quite as much reverence by the pair, however. In fact, in only a matter of years, the Brelands would end up playing a decisive role in the downfall of Skinner's reign with their work related to animal superstitions, identified a few years earlier by Skinner. During their animal-training regimes they occasionally noticed that some animals in their care responded to their training routines in ways that were odd and not at all expected. Animals were products of experience, sure, but every now and then something else was coming in to pull the strings that drive behaviours, they realised. Something unforeseen. Nurture couldn't explain the unusual behaviours they were starting to observe, they realised ... but nature might. When the Brelands plucked up the courage to publish their findings, Skinner's grip on psychology would decline.

Once more, as between Pavlov and Skinner, batons would pass. They always do. And with them, new characters would sprint onwards – this time, the cognition scientists. The next chapter charts their rise.

---

* The PDSA's advice on using clicker training for your dog can be found on the PDSA website: www.pdsa.org.uk.

CHAPTER 7

The cognition ignition

> 'What we see depends mainly on what we look for.'
>
> – John Lubbock (1834–1913)

Sure, to look at, Seleno wasn't much of a robot, but then neither was she much of a dog. Seleno was more of a box – a cube on three wheels with two light detectors (made using selenium) attached to the front. By modern standards, there was a hint of *Doctor Who* baddie about her, but audiences in the early twentieth century were enthralled. No one had ever seen anything like her. This was a contraption able to navigate, one metre a second, towards light sources (a torch) without any need for a caring thought. But Seleno had no brain and obviously could not think. Yet here she was, performing actions that until then were considered only possible for those in the realm of the living.

In fact, so jarring and unnerving was Seleno's efficient knack for finding her target that her inventors were horrified to imagine how the technology could one day, with a bit of tinkering, turn on its master. Seleno's co-inventor B. F. Miessner wrote: 'The electric dog which is now but an uncanny scientific curiosity may within the very near future become in truth a real "dog of war", without fear, without heart, without the human element so often susceptible to trickery, with but one purpose: to overtake and slay whatever comes within range of its senses at the will of its master.'

Thankfully this did not come to pass. Instead, in the modern day, we have now A1. A1 has no distinct neck or head region, but there are four legs and they are attached to a torso of some description, so there is no denying the hefty hint of dog about it. A1's legs whir like clockwork when it trots around. They move fluidly, forward-backward-forward-backward, propelled by a flurry of tiny pistons that pump in and out like tiny fingers working a flute.

In the promotional video, A1 certainly has the pace of a dog. It trundles along as if on invisible tramlines next to its human owner. The physics of it seem a little off, though. Like badly generated computer graphics, it almost seems to 'ghost' along the concrete. Rarely does it seem to ever accelerate, for instance. It just goes instantly between settings – slow, medium, fast. It just … is … at speed. This is the second thing that makes A1 most un-animal-like. The first is that – without teeth, perky ears and eyebrows – it is hard to ascribe to it anything by way of personality.

It really does move well, though. In the marketing footage, one impressive moment sees A1 zip along the pavement before coming upon what appears to be an impassable object. A1 is being approached by a human passer-by. An incoming human leg is in its way. A1 quietly considers this object with its camera eyes before moving to the left and circling around said object with the kind of complicated intensity a child riding a bike for the first time would do if faced with a giant inflatable crocodile in the road. An adult would navigate the unusual inflatable item effortlessly, of course, but a child would really take time to walk its way around the strange object. Moving around the crocodile would involve a *process* for the child. It would take negotiation. It's the same with the robot and the approaching human leg. It steers around it, sure, but it is not effortless. Far from it. A1 is clearly taking in and measuring up numerous variables that we (adult) humans take for granted.

A1 is eye-catching in the most vivid way. When it walks around, you can't really keep your eyes off it. As well as legs, the marketing videos take great delight in showing it, in slow-motion, navigating other day-to-day things with military precision. At one point the robot runs off a 5-inch pavement and manages not to tumble into a heap, for instance. As its legs clatter uncomfortably on the floor for a moment, it rights its centre of gravity and carries on walking as if nothing had really happened. It maintains pinpoint focus on its owner, throughout.

A1 is an animal-material hybrid or, to put it another way, an animat. A very expensive animat, made by Unitree, rival to US robotics company Boston Dynamics. A1 is also our avenue for thinking about brains and how they work – and how scientific understanding of the brain changed during the 1960s as Skinner's ideas began their gradual decline. Animats are, in essence, artificial animals. They can exist in physical creations (such as A1) and virtual simulations that play out on computer screens. I have always had a deep fascination with them, which stems from a childhood marvelling at contraptions like these on *Tomorrow's World* and later (don't judge me) on *Robot Wars*.

The animats I remember best were robot crickets. Robot crickets manage what many real insects achieve year in year out: in a busy world of interacting organisms, they manage with apparent effortlessness to locate one another. In the case of insects, this meeting has a purpose: sex. In the case of robot animat crickets, there is no purpose except to prove to scientists that it can be done, courtesy of some relatively simple coding.

To find one another in the wild, crickets depend on calls, but calls like these bring with them a challenge. Because crickets often share their habitat with lots of other cricket and grasshopper species, a given species needs to be adept at tuning in to the exact type of call that their kind emits. In other words, the call has to be just right for the cricket to begin making its advances.

It is the cricket's auditory anatomy that tells the insect to act upon the right kind of mating call. Sounds arriving into the body arrive at sensory neurons via two different inputs: one directly through the air and the other through a little tube on each side of the body. The length of this tube is very important – if the right kind of cricket call is incoming, the sound waves from the two inputs line up as if hitting a musical harmony. This is the call that, literally, excites them. Through this piece of physiology, cricket species become adapted to filtering out one another's calls from the mêlée of noises and sounds that spread across forest floors on warm spring days.

Once the cricket has registered a call from its own species, it begins to follow two very simple rules. The first rule is that, if the call is coming from the right-hand side of the cricket, the cricket should orient its body towards the right and continue to walk forwards. The second rule is that, if a call is coming from the left-hand side of the cricket, the cricket should orient its body towards the left and continue forwards. The two rules self-correct one another and, by doing so, the cricket moves closer and closer towards the source of the call. (Though greatly simplified, Seleno the electric dog worked on the same premise; it picked up light rather than sound.) In this way, a feedback loop occurs, much like that which occurs in a house thermostat: too cold and the boiler switches on; too hot and it switches off. In the same way that your house remains toasty, the lusty cricket finds its mate and Seleno finds, well, fame.

It really isn't too great a stretch to create an animat that will behave like a real cricket. Not many ingredients are needed: a motor, two wheels, microphones for ears, a microprocessor and a hundred or so lines of code for a brain. And the coding is actually very simple. This isn't to belittle the life experience of crickets, of course. My aim with this example is simply to underline that many of their day-to-day behaviours can be explained very easily by feedback loops – in other words, in the language of computers rather than Behaviourism.

The idea of simulating animal minds, particularly when it comes to exploring how their brains relate to behaviours, came in part from the work of Norbert Wiener, the American mathematician, philosopher and coiner of the term 'cybernetics' in 1948 – what he called 'the scientific study of control and communication in the animal and the machine'. From his earliest days, Wiener was clearly something of a genius. He graduated high school aged 11, began his zoological studies at Harvard aged 14, transferred to philosophy the following year, and had finished his undergraduate studies by the time you or I would have stumbled out of freshers' week.

In the context of this chapter, Wiener achieved two things: first, he invented a formal notion of what exactly a feedback loop was – an idea that could be applied to engineering, computer science, philosophy and even to the organisation of society; second, he looked at animals and considered that their brains may also be influenced by the same processes. To Wiener, what he saw in animal behaviours was because of feedback loops – hundreds of them. Thousands. Perhaps millions. He saw numbers. Almost, but not quite, circuits. Where Skinner was interested in notions of reinforcement and conditioning in animals, Wiener's ideas played to newly emerging faculties of science that favoured equations – the mathematicians, the theorisers, the physicists. These sciences opened up the idea that, inside the brains of animals, there was something that could be broken down into a series of connections no more complicated than feedback loops. His ideas found extra favour in a society making great strides in technological achievements after the Second World War – in computing, in weapons, in communications, in chemistry. No faculty, not even the brain, was outside the scope of this post-War wave of technological innovation.

Wiener's concepts melded neatly alongside the notions of other influential thinkers at the time, many of whom were also interested in how brains may mirror the actions of computers and vice versa. This includes John von Neumann (1903–57), the American-Hungarian mathematician, physicist and early computer pioneer whose conceptual 'stored programme generator' lives on in the computers, tablets and mobile phones we use today. It also includes Alan Turing (1912–54), the 'father of artificial intelligence' whose code-breaking exploits helped bring an end to the Second World War.

In time, this maelstrom of ideas would inspire other scientists and psychologists, each of whom would move them forwards in their own special ways. The notion of a brain as a machine was now back in vogue. The so-called 'cognitive movement' was rumbling.

'My problem is that I have been persecuted by an integer,' wrote the American psychologist George Armitage Miller in one of the finest ledes to a scientific paper ever written. 'For seven years this number has followed me around, has intruded in my most private data, and has assaulted me from the pages of our most public journals.'

Miller's 1956 essay about the number seven has become one of psychology's most cited works for two reasons. First, Miller observed a pattern in the ability of humans to remember things. He noticed that humans, when listing out things – names, cities, events – were drawn again and again to the number seven. Seven seas, seven deadly sins, the seven ages of man; that there are seven notes on a musical scale, seven colours in a rainbow, seven dwarves and so on.* He identified that humans can remember lists of seven things very well. But lists of fifteen or twenty-three things (or anything above seven) were often a stretch too far. Perhaps the brain wasn't an amorphous gloop of reinforced behaviours, after all, Miller deduced from this. Maybe, just maybe, brains had limits somehow 'baked in' through the language of genetics.

In considering minds in this way, Miller's views were aligning with a newly blossoming side-branch of animal science: the ethologists who, for two decades, had been considering animals in a similar vein. Just like Miller and Wiener, the ethologists were also expanding their range during the 1950s. Popularised by Dutch biologist Nikolaas Tinbergen and Austrians Konrad Lorenz and Karl von Frisch, ethology aimed to examine the behavioural processes in animals of numerous species and investigate how they came to be. The ethological approach involved both laboratory and field studies

* Miller also pointed to such everyday things as phone numbers – the fact that the digits in phone numbers are often split into fours and sevens, for instance.

and depended on a close alliance with evolutionary biologists, ecologists and, naturally, anatomists with an interest in brains.*

Ethologists like these were uncovering incredible things about animal instincts. There was the phenomenon of 'imprinting', where a newly hatched gosling displays immense interest and a dependence on the first thing it sees. How gull chicks are drawn to pecking at red spots on their parents' beak to encourage them to regurgitate food. How male stickleback fish, during spring breeding, show an inherent hatred of red things that happen to resemble the red belly of a potential rival. How bees intuitively 'waggle-dance' with one another in the nest, to communicate the location of faraway nectar sources. All of these examples hinted at a hidden call sheet of evolved behaviours.

Like these early ethologists, Miller saw that the brain's workings could be unpicked through careful application of the scientific method. That studies of mental processes were not as futile as other psychologists had deemed them. That there was more than just prior experience (reinforced actions based on positive and negative stimuli) to animal behaviours. That, perhaps, each and every brain, bestowed upon each and every animal alive today, comes ready-made with a suite of unique and interesting cognitive 'patches' or 'seats of specialism'. The psychologists of the time, most of whom were subscribers to Skinner's approach, sat up and took note.

Though he was trained in the school of Behaviourism, Miller became one of the most prominent researchers and theorists to challenge Skinner in the years that followed. In

---

* Tinbergen, Lorenz and von Frisch are considered by most to be the founding fathers of ethology, but they weren't. At the turn of the century, Turner – specifically his investigations into the innate intelligence of ants, spiders and wasps – was clearly taking an ethological approach. The same can be argued for Darwin.

1960, he opened his own research laboratory at Harvard and called it 'The Center of Cognitive Studies' – a deliberately antagonistic name.* But this new wave of enquiry would take time to bed in. The grip of Behaviourism was still strong. It was into this war of ideas that the Brelands planted their bomb – their finding that would cause Behaviourism to shake like never before, and tip the scales towards the cognitive sciences in the decades that followed. Here, briefly, are the details.

During their professional work with Animal Behavior Enterprises, the Brelands had occasionally noticed something strange about the animals in their care: they noted a number of cases of unusual animal behaviour that appeared unexplainable using the reductionist lingo instilled in them by Skinner. It began with chickens.

In one of their commercial endeavours in the 1950s, Keller and Marian Breland had trained some chickens to move towards a makeshift novelty jukebox, pull a rubber loop to start music playing and then strut over to a slightly raised disc to begin scratching and kicking the floor around and around in a manner to look like the chicken was dancing. At this, the chickens duly obliged. The problem that came later, however, was that when they tried to train the same chickens to do something else, like standing still on the platform and posing for the camera, half of the chickens trained could not manage it. They couldn't be made to stop pecking and scratching. It seemed these chickens had developed an irrepressible habit, unlocked in them by the first trial, which was a standard behaviour that chickens exhibit when allowed to fend for themselves.

* Miller later wrote: 'For someone raised to respect reductionist science, "cognitive psychology" made a definite statement. It meant that I was interested in the mind.'

No doubt, fans of Skinner would have pointed to this being an example of animal superstition. So, considering this, the Brelands noted down this strange phenomenon and carried on with their training regimes.

The next strange occurrence happened with raccoons being trained to pick up coins and post them into metal containers. 'Raccoons condition readily, have good appetites, and this one was quite tame and an eager subject,' the Brelands stated in their research report. 'We anticipated no trouble.'

Conditioning the raccoon to pick up the first coin proved to be simple – raccoons have rather primate-like paws, after all. But this led to the first problem: the raccoon refused to let go of the coin. The second, far worse, problem was that the raccoon would often begin rubbing the coin against his body in a most rigorous and semi-erotic manner. Sometimes the raccoon did almost manage to post the coin successfully, flirting with the coin slot but then – curses! – it would pull the coin out once more for another round of energetic body-rubbing. Posting one coin proved difficult for the raccoon, but adding another coin brought new problems: 'Now the raccoon really had problems (and so did we). Not only could he not let go of the coins, but he spent seconds, even minutes, rubbing them together (in a most miserly fashion), and dipping them into the container,' the pair observed.

Worse still, in spite of the Brelands' best efforts, the raccoon (and others like it) actually seemed to be doing it more and more – the coin-rubbing behaviour was sticking and somehow positively reinforcing itself within the raccoons in their care. It goes without saying that the raccoons rubbed themselves out of the job. In the end, the Brelands had to cancel their plans and inform their client (a well-known high-street bank) that their dream of a raccoon posting money into a novelty piggy bank could not be realised. (They went with raccoons throwing basketballs into little hoops instead.)

In their article, Keller and Marian Breland recounted other strange anecdotes about the animals in their care: pigs that couldn't post large wooden coins into other novelty piggy banks without dropping them and rolling them around in the mud beforehand; cows that couldn't be trained to kick; more rogue chickens who refused to be told how and on which plinth to dance. 'These egregious failures came as a rather considerable shock to us, for there was nothing in our background in behaviorism to prepare us for such gross inabilities to predict and control the behavior of animals with which we had been working for years,' they summed up.

And then, in 1961, they published 'The Misbehavior of Organisms'. The title was a direct reference to Skinner's famous 1938 book, *The Behavior of Organisms*, and a knowing nod to those psychologists becoming frustrated at Skinner's hold on psychology and animal behaviour. It certainly garnered plenty of attention and support from those who considered animal brains as more than just a *tabula rasa* upon which conditioning acts equally across species.*

The Brelands labelled the phenomenon they saw 'Instinctive Drift' – when animals revert to innate behaviours (of nature) that interfere with conditioned responses (of nurture).

In ways you are likely to know, dogs express many innate behaviours that resemble those of the washing raccoons. It is no surprise that many of the things we train dogs to do best

* The Brelands never fully turned their back on Behaviourism in later years. In an interview with animal behaviourist and veterinarian Sophia Yin in 2001, Marian Breland restated her intentions for publishing 'The Misbehavior of Organisms': 'it didn't mean that we were abandoning operant conditioning or behavior analysis at all; it just meant that people needed to look at these other animals and see what they're doing because they had forgotten how to do that.' Apparently, Skinner and the Brelands made up and remained friends.

are the things they expressed in their wild ancestry: to chase and retrieve items, to stay still, to stare with intensity at something, to herd (stalk). The things you see in your dog aren't written in the stars; they are the products of both positive and negative reinforcements, sure, but they are more than this. They are the products of genes too.

As the 1960s progressed, Skinner's hardline stance on animal behaviour started to look more and more shaky. The Brelands' paper and Miller's essay were just the start. Other cognition scientists would follow, each dropping ideas and observations that fell like heavy stones onto the grave of radical Behaviourism. Particularly notable was John O'Keefe, the American-British neuroscientist who would go on to win the Nobel Prize in Physiology or Medicine for his findings. A decade after the Brelands published, O'Keefe's research added something of a final nail in the coffin of Behaviourism. By applying electrodes to a brain region called the hippocampus, O'Keefe and colleagues used rats to discover that brain activity spiked not from particular sights or sounds, but rather from the rats being in a particular part of their enclosure. O'Keefe helped discover what are called 'place cells' – the hippocampal neurons that animals use to map their surroundings.

In the decades that followed, more discoveries like this would be made. Particularly incredible was the discovery that some brain cells fire depending on the particular direction that an animal subject is orientated towards. And then there was the discovery of strange cells that light up in the entorhinal cortex (a clutch of neurons closely aligned with the hippocampus). These cells do so in discrete patterns that match up against distances paced out in the real world, even down to 30cm. Discoveries like these showed how, through the links between the entorhinal cortex and the hippocampus, the brain was forming a mental representation of space. Animals had brains hard-wired to map their surroundings.

'It might not *look* like a conventional map because it's not written on parchment and isn't labelled with printed text and a compass rose,' writes the behavioural neuroscientist Kate Jeffery in her wonderful article 'Maps in the Head'. 'However, the neurons in these regions respond in a way that shows that they are somehow stimulated, not by bells and food, as the Behaviourists believed, but by abstract properties of the animal's experience, such as how far it has walked and what place it has reached'.

The radical behaviourists were dead in the water by the time these later discoveries about place cells and mental maps were made. No level of stimulus-response associations or operant conditioning could produce such a structure. The brain was hardwired from birth, it seemed, to understand unique principles. It came ready-made, gifted through fortuitous lines that dodged the fateful hand of extinction. The logic of reductionism that had proved inescapable for almost fifty years gave way to something newer and in many ways far more exciting. Animals weren't only the product of their experiences. They were also, in a very real and physical sense, born that way.

In their own way, Keller and Marian Breland and George Miller were – like Wiener – pushing back against the mainstream ideas of psychology and biology, opening the doors for geneticists, the ethologists, the ecologists, the neuroscientists and the mathematicians in their wake. A brave new academic frontier was calling. A new world where the temperaments and behaviours of animals were products of experience, but also dynamic and varied products of evolution – of genetics, of inheritance, of fortune. Of both nature and nurture. Into this maelstrom of ideas, edged the dogs.

CHAPTER 8

How nature met nurture

> 'Experience is augmented, facts appear which do not agree with it, and one is forced to go in search of a new mode of conceptualization within which these facts can also be accommodated; and in this manner, no doubt, modes of conceptualization will be altered from age to age, as experience is broadened, and the complete truth may perhaps never be attained.'
>
> – Jöns Jacob Berzelius (1779–1848)

Though you won't know them by name, your life will have been touched by John Paul Scott (1909–2000) and John Langworthy Fuller (1910–92). Your connection with these men happened when you reached for that handbook on raising puppies or googled the developmental milestones you should expect in your dog. The first opening of the eyes. The first play-fights, the first forays away from mother, the first signs of weaning, the first signs of going it alone. You'll know your path has crossed these two men's when you consider moments like these – when you socialise your puppy with other dogs, or let them have experience with loud motorbikes and bin lorries, with kids at the park or cats in your mother's house, or when you let your puppy mix or mingle and do as much as possible of the day-to-day things. Mostly, you have these men to thank for knowing that this is a very good thing for your puppy.

It was Scott and Fuller's research that changed how we raise dogs today, although it is research that has been almost totally forgotten by many modern dog-lovers, veterinarians and even some dog researchers. Their quiet, ethereal influence is further underlined by the fact that the pair have a combined internet footprint comparable to that of a third-rate 1960s footballer. One of the rare photos that exist online sees J. P. Scott playing beyond typecast: all white coats, arms folded, glasses – relatively smiley and forthcoming but suitably reserved.

The thing that set Scott apart from other researchers was that he refused to focus his interest in animal behaviour on just one clutch of species. As well as dogs, Scott's research animals included sheep, goats and house mice. But Scott, like many of his time, had seen two World Wars and a Depression and this too had influenced his outlook on life. He became interested in how development affects behaviour. From the start, he saw dogs as a potential research animal to help pull apart and expose how small influences in infancy affect the personality of human adults. What he lacked was an institution with the guts (and financial clout) to fund a long-term project that could measure animals and their development over years and decades rather than weeks and months. That was until 1945, when the Jackson Laboratory came calling.

At that time, the Jackson Laboratory in Bar Harbor (Maine) had recently been endowed with a large financial grant from the Rockefeller Foundation, who had become very keen to fund projects exploring genetics and behaviour. Being the foremost geneticist (actually, pretty much the *only* geneticist) in the country, Scott was, in his own words 'the logical choice for heading up this new program'. Originally trained as a physiological ecologist, Fuller was employed soon after. What started as an interest became a fascination and, within months, he switched silos and a new career in behavioural genetics was set alongside Scott.*

Scott and Fuller were the first to compare different dogs under environmentally identical conditions and to measure behavioural differences between breeds – identifying and isolating the influence of nature (heritable traits) and nurture (experience) upon individuals. Key among their discoveries was that the life experience of dogs mattered as much as the genetics and that, only when both aspects of a dog were considered, could owners expect the very best behaviours. They encouraged dog-friendly training practices. They documented 'hybrid vigour' – how first-generation crosses of breeds often resulted in dogs that performed better in problem-solving situations, for instance. They showed the emotional benefits of puppies staying with their mother for longer and how damaging in later life early separation could be. They observed and recorded posture. Investigative

* A rare recording of Scott (in the Jackson Laboratory Historical Archives) has him suggest that he was especially keen for Fuller's involvement, partly to allay his own worry that if something untoward would happen – for instance, if Scott were to pass away suddenly from a heart attack – the project (and funding from Rockefeller) could continue in someone else's safe hands.

behaviour. Vocalisation. The docility of dogs across breeds across different life stages. Their aggression and their play. In short, they wrote the manual on the life events that shape the personalities of adult dogs.

The pair focused their efforts on five dog breeds. There were beagles, wire-haired fox terriers, Shetland sheepdogs, American cocker spaniels and the African basenji, then considered a breed most reflective of ancestral dog-kind. Key to the experimental set-up was that, when it came to raising the hundreds of puppies used in the trials, each puppy of every breed was to be raised in an identical manner. By doing this, it was hoped that the trials would reveal behaviours caused by genetics rather than by the environment. As the puppies grew, each was given a standard set of behavioural tests that looked at things like play behaviours, how readily puppies could be picked up and cuddled and how easily they could be trained to undertake simple commands like walking to heel. The cognitive abilities of each breed were also investigated, through the use of 'mazes' in enclosures or tests such as seeing if dogs could pull a bowl of food from underneath a wire mesh.

The abilities of some breeds in their experiments were quickly shown to be genetic. For instance, beagles could – quite predictably – use their nose to locate a mouse in a novel enclosure within sixty seconds of being introduced to it. Scottish terriers, on the other hand, performed abysmally; one report has it that a Scottie inadvertently trod on a mouse at one point without even realising it was there.

To Scott and Fuller, it was no surprise that such a sensory prowess could be under the influence of genetics. After all, beagles had been bred for this very purpose for centuries. However, what was more of a surprise came next. The received wisdom at the time was that different breeds had quite different temperaments (or to put it another way, 'personalities'), endowed upon their recent ancestors through years of selective breeding. Yet, incredibly, the experiments that Scott and Fuller undertook began to suggest the opposite: that 'nurture' shaped personality as much as – or indeed more than – nature. While some breeds did show certain behavioural leanings (cocker spaniel puppies, for instance, showed less inclination to play), on the whole, all the dog breeds were much of a muchness. In fact, the pair discovered that behavioural traits often differed more within a breed than between them, with males and females differing in their behaviours far more than breeds. Only the basenjis showed a consistent difference in some aspects of their behaviour, preferring not to be handled until five weeks of age, for instance.

A key finding of the experiments was one that we now accept with barely a thought: that puppies have a clearly defined sensitive period in their development, in which the personality of the adult dog is shaped. Scott split this early period into three: first there was the neonatal period (in which nursing is established); then the transition period (in which sensory processes are sharpened and puppies begin to move around); then follows the socialisation period (the period between four and twelve weeks when social relations and attachments form).*

To reach this important conclusion, the team observed puppies through one-way glass in ten-minute intervals at periods throughout their first sixteen weeks, noting down all aspects of their behaviour. The pair also began isolation tests – removing puppies temporarily from their mother and siblings to observe their responses. Their results showed how, from four weeks to sixteen weeks, puppies find it incredibly easy to adjust to new people, animals and situations. Conversely, they showed how, if puppies are denied this period, their personalities are impacted negatively. They become fearful, unpredictable – potentially dangerous. They

---

* This isn't to say that, by the time they are twelve weeks old, dogs are good to go. The juvenile period (often considered to last until sexual maturity) continues to be an important time for socialisation, second only to the initial development phases that Scott and Fuller outlined.

defined this critical period as 'a special time in life when a small amount of experience will produce a great effect on later behavior'. They used the analogy of a high-powered rifle: a small push on the trigger is all it takes to cause great damage a distance away.

Scott and Fuller's laboratory team also investigated puppies given limited contact with humans. What resulted were fearful, timid dogs that preferentially chose the company of other dogs rather than the company of humans. This discovery, perhaps more than any other, had enormous significance for the ethical treatment of dogs in experimental research. It meant that laboratory-raised dogs, reared away from regular contact with humans and other dogs, were at higher risk of having some degree of emotional damage. This single long-term experiment would forever change the fortunes of dogs used in research.

In many ways, the discovery of a critical period* matched what influential ethologists such as Konrad Lorenz had found in their observations of animals in the wild and in the laboratory. Famously, Lorenz had discovered that greylag goose chicks exposed very early in life to his presence quickly learned to follow him around, acting as if he were their mother. Crucially, he determined that this 'imprinting' would only occur in a distinct time period of a few hours – the so-called 'critical period' of development.†

Scott and Fuller put their findings into the 1965 book *Genetics and the Social Behavior of the Dog* – something of a bestseller, shipping millions of copies in the decades that

---

* In the modern day, most researchers favour the term 'sensitive periods' to allow wriggle room in the onsets and offsets of different and less well-defined sensitivities.

† Many researchers point out that imprinting (an idea popularly associated with Lorenz) was actually discovered by the British biologist, Douglas Spalding (1841–77).

would follow. Its success was a surprise to the authors: 'The book was well received by scientists, and to our surprise by serious breeders and trainers of dogs,' wrote Fuller in 1989. Yet, considering the years of laboratory study that went into its creation and the fact that so many breeders have a copy on their shelf, the book is still enormously under-appreciated as a treatise in how what we know about dogs comes from two decades of hard-won science rather than flaky received wisdoms. The closest it came to winning an award was when it was named 'the best dog book of 1965' by The Dog Writers of America.

In the modern day, due to the book's influence, dog welfare organisations and breeding organisations recommend as much structured socialisation with other dogs as possible.* They also encourage repeated (and controlled) encounters with, among other things, children, cats, horses, livestock, cars and motorbikes.

In more practical terms, Scott and Fuller's study influenced the way that guide dogs for the visually impaired are trained. There were other benefits too – in particular their research facility, which was an important resource for visiting scientists eager to learn from two of the country's foremost geneticists. In archive recordings, Scott, without a hint of false modesty, states that almost every single scientist who would go on to make their name in the fields of comparative psychology and

* The word 'structured' is important here: 'puppy parties' (where lots of puppies are brought together in a kind of 'free-for-all'), for instance, were once considered good practice, but this advice is changing. 'For nervous puppies, they can be overwhelming, escalating their fear and potentially doing more harm than good,' Sean Wensley, a senior vet at the PDSA, informs me. 'Current advice is about practices offering sessions that provide structured, supervised interactions which assess and accommodate the needs of individuals, rather than simply a group meeting and playing with each other.' The PDSA's advice is available on their website, www.pdsa.org.uk.

animal behaviour would pass through its doors, either as casual visitors, summer investigators, research assistants or graduate assistants. Not many research institutes can boast such lofty claims as this. Yet the impact the facility had in inspiring new generations has largely been forgotten. When towering on the shoulders of giants, it becomes harder and harder to see the outstanding features and achievements of those below, I guess.

'So much of what I've done has become common knowledge,' Scott said in 1994, six years before his passing. 'No one realizes someone had to document when puppies first open their eyes. It's actually gratifying that so many are familiar with my results – even if they don't know how they came about.'

Part of the reason the pair's contribution to science is not wider known is undoubtedly down to their own personal natures. The fleeting references to them that still exist depict them as rather quiet, lacking showiness, empathic and fairly good-natured. Scott had his own reasons for wanting to focus on dogs. ('There is no friend as loyal as a dog,' is a line for which he is credited.) But, importantly, dogs also helped Scott continue his side interest in human and non-human aggression, an emotional response he considered to be the product of complex interactions between both nature and nurture. His belief was that criminals weren't born bad; they were made bad by circumstance.

Dogs helped Scott explore this. In fact, some of the experiments – including keeping a subset of dogs in varying degrees of solitary confinement and assessing the impact on development – were clearly designed to explore the parallels between humans and dogs. This fact is given away by the informal name of one particular dog experiment involving raising dogs in isolation – it was known as the 'Kaspar Hauser' study, named after a German child said to have been kept in a darkened cell for the first ten years of his life.

'Scott's research interests were always informed by the great world events of his time and his concern with conducting research that might reduce violence, foster peace, and improve economic prosperity,' wrote the psychologist Donald A. Dewsbury in a touching obituary. By all accounts, Scott seems quiet, focused, driven and lacking pizazz. His social conscience rings loud. He sounds quite likeable, really.

Unlike Scott, Fuller seems to have barely received an obituary at all, not even in his local newspaper. He did, later in his life, write the following about himself, after rereading his old research reports on dogs: 'Yes, I recalled the experiments but the details of procedure and the discussions might have been the work of another person. The redeeming feature of this lapse of memory was that I liked what this alter ego had written several decades ago. He was intelligible, not too long-winded, and I agreed with most of his conclusions.' The irony of one of the world's leading developmental geneticists gently critiquing his earlier work should not be lost on any of us.

Still, no one can doubt the influence that both men had on modern-day best practice. In my personal experience, not a day goes by when I am not reminded of the need for socialisation – it's become a guiding principle for dog welfare across the world. 'It's among the most important advice we ever impart for dog welfare,' Sean Wensley, senior vet at the PDSA, tells me. 'We advise on it emphatically, whenever we can – both for people selecting where to acquire a dog, so that, for instance, the breeder has adequately socialised their pups; and at the first vaccination, so that the good work continues by the new owner for the remaining critical weeks.'

Niki Khan, reproductive ecologist at Nottingham Trent University, agrees: 'In a nutshell, Scott and Fuller's Bar Harbor experiments were groundbreaking,' she informs me. 'Although these studies are now more than sixty years old, their results are still relevant today.'

Notably, they inspired whole learn lines of research, many of which have only recently begun to bear fruit. 'These vary widely, involving behaviour genetics, maternal care, the early developmental environment, and the transgenerational effects of maternal hormones, in animals ranging from rodents to birds and, of course, primates,' Khan explains.

Together Scott and Fuller's research showed that how we treat puppies mattered. That personality could be cemented in just a few early moments of life. But, significantly, their research allowed us to apply this knowledge to humans, too. The pair could clearly see the potential for applying their learning to parents, to schools and to wider human society. Dogs helped the pair reach a novel perspective that would go on to influence psychology, in particular, for decades.

But there were other benefits too. In writing *Genetics and the Social Behavior of the Dog*, the pair would also go some way to resurrecting the authenticity of behavioural genetics within wider biology. In previous decades, such research had garnered a problematic reputation – particularly by being associated with eugenics – but Scott and Fuller helped bring back the study of inherited character traits in an altogether more academic, rigorous and ethically sensitive manner.

Through the twenty-year Bar Harbor study, dogs found a new role in science – they could help us learn about ourselves, about our experiences, our reality, our moods and behaviours. In more ways than one, through them, we learned more about the human condition. This was quite the innovative perspective at the time and so, naturally, the psychologists were interested. Very interested. For, at about the same time as Scott and Fuller's research was playing out, psychologists had been cultivating their own ideas about dogs and how their behaviours could mirror those of humans. Importantly, they saw in dogs a potential way to explore other aspects of the human condition. And, still more importantly, they saw in dogs a way to cure, to grow and flourish, and to make the world a better place – for both species.

There was a darker side to Pavlov's notion of conditioning: if animals could be conditioned with reward, they could also be conditioned through punishment, pain and suffering. Animals could be made to do things in order to reduce their own misery. To save their own skins. Suitably ghoulish and (to our modern eyes) almost unspeakably cruel, this area of study had a name: aversive conditioning. Aversive conditioning was yet another very 1950s interest: how animals could, through the use of tones or lights, come to associate, anticipate and avoid negative stimuli, often in the form of that other 1950s go-to solution: electrical shocks. It was into this ethically troubling arena that a young Martin Seligman was to begin his career in psychology. You might not immediately recognise the name Martin Seligman, but it is likely that you know the treatment he later helped popularise: Cognitive Behavioural Therapy, or CBT. First, however, Seligman would need his 'Eureka!' moment, and this unique research environment would provide it.

It was during this early research that Seligman discovered the phenomenon of 'learned helplessness' in dogs, a kind of emotional malaise where dogs resist the urge to fight. They endure and suffer, shutting themselves off to finding solutions in intensely stressful situations.* They give up, in other words.

To understand learned helplessness, it helps to spend a few moments covering the experimental methodology under

* I have chosen to leave many of the details of Seligman's initial experiments involving electrical shocks out of the main text of this chapter – it reads too bleak. Those interested can find the details in the 'Research Notes and Further Reading' section at the end.

which Seligman (alongside psychologist Steven Maier) worked. Together, the research team utilised an experimental set-up favoured by others in his laboratory – the use of 'shuttle-boxes' (so called because subjects can 'shuttle' from one side to another to avoid negative stimuli). A typical shuttle-box set-up involves a box being split into two sections by a little raised wall over which the dog can easily vault to safety. In a shuttle-box, to get away from a shock, a dog can move to the comparative safety of the other side of the box. Yet some dogs did not do this. Quite to the surprise of psychologists (and, indeed, everyone involved), some of the dogs instead sat still when electric shocks were administered; they chose to endure the electricity without apparently fighting back or vaulting to the other side. They seemed, for reasons unknown, to suffer their fate. Seligman's mind was set whirring. A decade later, he would publish.*

Looking back now, it is hard to imagine how laboratory operatives could see dogs suffering in such a way. But cruelty for cruelty's sake was never Seligman's intention. Significantly, Seligman saw in these dogs a way to investigate human mental illness – he saw a clinical sign of depression in them, in other words. He also understood that traumatic events, such as those endured by his study animals, were capable of shaping their subsequent learning behaviours, their emotions, the motivations – their outlook on life, essentially. And, crucially, he sought to somehow 'repair' the dogs showing learned helplessness. He wanted to trial approaches to put them right, to reduce the damage he had caused his research animals. And so, like a clockmaker pulling apart a watch in

* A quick aside: fifty years after first coining the term 'learned helplessness', Seligman and Maier are no longer convinced the term best describes the phenomenon: 'Passivity in response to shock is not learned. It is the default, unlearned response to prolonged aversive events,' they write.

order to understand how to put it back together again, Seligman set about trying to repair the mental states of his dogs after their traumatic suffering at his hands.

First, this was attempted manually. When the electric shocks came, he and his colleagues physically moved the helpless dogs around their shuttle-boxes in an attempt to show them a solution. This sounds simple but it worked: it had some success in reducing rates of learned helplessness. Other approaches also proved helpful in 'curing' learned helplessness. A key moment was discovering that learned helplessness could be 'inoculated' against by giving dogs experience of a lever, before the true experiments began, that stopped electric shocks. Naturally, Seligman looked at his research and how readily it could be applied to humans. If learned helplessness could be reversed in dogs, then might this be the case in humans? It was a gallant question, particularly for an era in which mental health still carried with it some stigma (as it still does in some quarters of society today). And the answer was yes: it could be applied to humans. The condition of learned helplessness can be treated.

Since its being put to paper by Seligman and Maier, learned helplessness has been found to factor in a wide range of human social situations. It presents itself in some emotionally abusive relationships, such as in cases of child abuse where learned helplessness becomes a survival strategy to shut down or numb experience; it is found in the classroom, when 'non-academic' students unsuited to a given learning style simply give up; it is known to the socially anxious, the poor, in minorities overlooked and shifted to the sidelines of society. Thankfully, there are now treatments that can offer some kind of solution. The most well known, as mentioned earlier, is CBT formally devised by Aaron Beck, with whom Seligman worked, at about the same time that Scott and Fuller were publishing their magnum opus on dogs.

The research that Seligman undertook on dogs was no flash in the pan. According to Maria Konnikova's fascinating 2015 article in the *New Yorker*, his research findings continued to influence psychology in the decades that followed – particularly when it comes to assessing whether, and to what degree, learned helplessness can be challenged, averted or turned around. Among Seligman's most famous studies was a longitudinal research project, undertaken alongside colleagues, of two schools in a suburb of Philadelphia, involving fifth- and sixth-graders. Over a period of three months, according to the study, students who had identified as having suffered depression previously, or were deemed at high risk of suffering depression, met for weekly therapy sessions. Key to these sessions was that the students identified their negative thoughts and pulled them apart to identify them objectively before considering alternative ways of thinking about them. Instead of thinking, perhaps 'my dad hates me and that's why he left', attendees explored other reasons, such as that a missing father was failing to deal with the pressures of work, addiction or his own unknown, undiagnosed trauma. Quite simply, the pupils drew out pessimism and turned to an alternative narrative with hope.

The results of this longitudinal study supported Seligman's finding with dogs: that humans could be inoculated against feelings of helplessness through a balanced reappraisal of a given situation. First, the children that completed the programme were clearly less depressed than the control group of students. Second, and far more spectacular, was that the difference between the two groups grew more pronounced as time went on. After the first year, 7 per cent of the students involved in the classes reported mild to severe depression, compared to 29 per cent in the control group. And after two years, this relationship continued: 44 per cent of the control group suffered depression by this point, yet the students involved in the study were at half that number. So successful was this study that it spawned a larger longitudinal project, the Penn Resiliency, that draws upon a larger sample size and has inspired sister projects in Australia and in the UK, specifically in South Tyneside, Hertfordshire and in Manchester.

Projects like these have at their core Seligman's original treatment for alleviating learned helplessness in dogs: that individuals, with help, can find a way to gain control of their lives in the face of uncontrollable trauma. By knowing there are instruments to find relief. By being reminded that there is a way out in tough times.

In the course of my professional work, through visiting shelters and interviewing those who work with abused dogs, I have seen first-hand the familiar stories of neglect. Most noticeable, almost every time, is the impact that past trauma has on the present lives of dogs: the dogs that barely have the energy to wag their tails; the dogs seemingly uninterested in new games, in socialising or in investigating new smells; the dogs that look like they want to bite; the ones that howl and whimper endlessly. It is visible in the opposites too: the dogs that sleep an inordinate amount of time; the dogs with bowls of food that once would have been devoured, now remaining wholly untouched. The learned helplessness of the learned helpless. The stories behind these dogs frequently overlap. Often they are incredibly disturbing. Nearly always, in the modern age, they are unnecessary.

If there is a message that rings out loud and clear from Scott and Fuller's work, it is that bad dogs aren't born bad. If there is a message that rings out from Seligman's discoveries, it is the bad dogs could be if not completely cured then made better – or better than they were, following the abuse or trauma, at least. But Scott and Fuller and

Seligman's research should also be remembered for something else: like some kind of Pandora's box being opened, it changed the relationship we could knowingly have with laboratory dogs.

By exposing the lasting emotional damage that experiments could impose upon dogs bred for the laboratory, this made our experimentation on them an increasingly difficult position to defend by the late 1960s. This, alongside ever-more-heated public outcry against vivisection experiments playing out in the USA, saw an interesting sea change begin to take place: it was no longer acceptable for dogs to be victims of science. At the very least, they would have to become collaborators instead. And so our research questions about them changed. Before, in the first half of the twentieth century, we had tugged hard on the metaphorical leash, keeping dogs repressed to quench our scientific curiosity. Now, in the second half, as we eased the restraints to limit the damage inflicted, we realised dogs could take us somewhere far more interesting. So we let dogs guide us and take us to new places. We loosened the leash.

This emancipation, of sorts, occurred as Behaviourism was washing out from within academia in the late 1960s, particularly in psychology. In its place, cognitive psychology – a wider empirical study of mental processes, including thinking – was occurring. One approach viewed personality through the prism of prior experience. The other approach saw the roots of personality in both nature and nurture, and was (the irony) salivating at the potential for new discoveries.

In this new age of scientific endeavour, with these new tools of scientific enquiry, dogs could be asked new and exciting questions in a more compassionate, empathic manner. And these questions were familiar. In fact, they were much like the questions asked by Darwin and friends a century beforehand: questions about experience, emotions, their capacity for consciousness. To think. To dream. To know.

Finally, mainstream science was ready to get into the mind of dogs in a manner that didn't involve sharpened instruments or torturous devices. In a manner that required us to think like them. Finally, a new age in the human–dog relationship could dawn – a scientific friendship, rekindled.

SECTION III

MEET, PLAY, LOVE

CHAPTER 9

What is it like to be a dog?

> 'How can we feel sure that an old dog with an excellent memory and some power of imagination, as shown by his dreams, never reflects on his past pleasures or pains in the chase?'
>
> – Charles Darwin, *The Descent of Man, and Selection in Relation to Sex* (1871)

As far as party pieces go at the end of a lecture, Don Griffin's was very hard to beat. TEDx would surely not have allowed such a stunt. Too much could go wrong. It could damage the lights or the camera equipment. It might cause a stampede. But in the 1950s barely anyone had worries like these. And probably no one cared much about the poor bat, whose job it was to fly around wowing the crowd. Always leave them wanting more, goes the adage. Griffin did that … with a bat.

Griffin – a rather preppie-looking, bespectacled professor – was something of an academic celebrity by then because he, alongside Robert Galambos, was co-discoverer of the incredible fact that bats used a form of sonar to navigate their surroundings in the dark. Bats emitted clicks, the pair discovered, that bounced off nearby objects and could be detected by the ears, allowing a bat to construct a three-dimensional map of its surroundings based on echoes. They called their discovery echolocation.

At first, the idea of echolocation had been laughed at by science but, by the 1950s, it had become not only accepted but positively lauded by zoologists. This was, after all, an entirely new way of taking in the world. Truly, theirs was a revolutionary discovery – one that changed our understanding of how animal sensory equipment could be rejigged for new purposes by the whittling force of natural selection.*

For the purpose of our story, one particularly significant demonstration by Griffin took place at Rockefeller University (New York City) in one of the dining rooms. At the end of his lecture Griffin released a single bat into the air so that the audience could marvel at the bat's otherworldly dynamism, negotiating objects almost completely without the use of sight. It must have been enthralling for everyone (except perhaps the kitchen staff whose job it was to clean up). Sitting in the audience during this presentation, gazing at the bat in a state of total wonder, was a young visiting philosopher by the name of Thomas Nagel. Just as Skinner had once watched Pavlov giving a lecture, Nagel was watching Griffin and the

* We now know that bats are not the only animals to navigate using such specialist skills. Other animals capable of using biosonar include dolphins, swiftlets, tenrecs and even shrews. Incredibly, some sight-impaired humans have shown a capacity for using the technique.

bat; he was awed and amazed by this new perspective on life. He watched the bat, rapt. He considered what it would be like to be that bat, flittering and flapping and twisting and turning about in the dining hall, its world calling back to him through the unimaginable medium of reflected sounds. The fact that Nagel could not put the bat's experience into words is what made the experience all the more beautiful. He ended up carrying the strange thought experiment with him. As Nagel's career developed, the experience with Griffin's bats stayed. Two decades later, he would try again to put the bat's experience into words – and fail in the most celebrated of ways.

Nagel's 1974 essay 'What Is It Like to Be a Bat?' posited a central immovable problem of this form of science: that there can be no objective criteria that can help us *know* truly what it feels like to be in the mind of another animal.* In the case of bats – living not in a world of colours or textures but in an environment made up of surfaces from which echoes emanate – it is almost impossible for us to imagine, being that our senses are heavily evolved towards visual stimuli, the most developed of our primate senses. Our brains just aren't wired that way. Even for us to draw in our mind's eye a world of echoes, we imagine a 'wall' of sound – a kind of visual playing field of rebounding sound waves. It is impossible for us to get away from, no matter how we try.

Nagel's essay didn't remove animals of their agency to feel. Many animals might feel emotions in stronger and more profound ways than we can know. Instead, his argument is that we'll never have a true understanding of what it must be like to be another animal because objective, reductionist

* In many ways, Nagel's argument aligned closely with some biologists, particularly Jakob Johann von Uexküll (1864–1944) who first coined the term '*umveldt*' to describe the perceptual world in which a given organism exists.

thinking about consciousness is impossible with subjective minds like those that animals (including humans) possess. Nagel argued that, while scientific investigation might chisel away at our doubts about consciousness, it can never get us to the 'facts beyond the reach of human concepts'. After publishing the essay, which would become something of a sacred text to philosophers,* the scientists and psychologists of the 1970s predictably divided into two warring camps: those who considered consciousness to be a viable area of study and those who certainly did not.

Does this mean we can never know what it is like to be a dog? Dogs are, in some way, not unlike bats. Like bats, their senses – the window through which their representation of the world is made – are skewed in a way that differs to our own, making our understanding their world that bit harder. Imagine, for a moment, being a dog and touring your kitchen by smell. Imagine pushing open the door and taking in the oily scent of the hinges. The oven gloves that, though washed many times, are still soiled by the smells of a hundred roast dinners. The cupboard in which the biscuits are kept. The tiny gap under the kitchen sideboard from which molecules leak out of a mouse that died two years ago and that no human ever knew was there. The smells that the houseplants release when humans brush past. The scent of the wet soil. The smell of dead leaves, of petrol, of wet patio slabs; the scent of ghostly night-time animals whose odours seep into the house via the gap in the window frame. The footprints of the child who visited last week. Where the food bowl goes. Where it has been. The exact location on the bowl where your beloved fingers have just gripped. If we could think like a dog, our daily lives would be taken up with tiny observations

* Daniel Dennett, the well-known American philosopher, calls Nagel's essay 'the most widely cited and influential thought experiment about consciousness'.

such as these. Stories would be told through smell. The vast majority of poems would be written about them. Our culture would feel very different indeed had we evolved from wolves rather than apes.

Even the act of sniffing in dogs belies an incredibly complex set of actions. In comparison, gathering light is rather a straightforward process. Our eyes are like satellite dishes that pick up its movement, absorbing a range of different wavelengths that penetrate and ricochet around in our atmosphere. Molecules of odour do not behave like light. They dance like unseen fairies upon tiny waves and eddies in the gaseous traffic jam within which our world plays out. They drift, they are diluted and they denature – decaying into other states, each with their own unique odours and their own evocations. To pick odour molecules out of the molecular maelstrom takes its own special hardware, which dogs keep quite literally front and centre.

Dogs manage their sense of smell through some evolutionary innovations that are unavailable to many other mammals. First, when dogs inhale air, a fold of tissue splits the incoming air into two streams. The vast majority of air heads to the lungs, providing oxygen for respiration, but the other channel – perhaps 10 per cent or so – is directed nearer to the brain, to a region packed with curly bones called turbinates. The turbinates look a little like the air filters in a car and they act in a similar way too. As air moves through these thin tissues it collides with millions of receptor sites, each primed to receive a different kind of scent molecule. Every time a scent molecule plugs into one of these receptors, a message pings to the brain: an odour is analysed and a smell registered.

It goes without saying that a dog's sense of smell is vastly superior to our own, but it is their ability to detect tiny traces of odours that particularly sets dogs apart from many other mammals. Some dog breeds have noses that may be 10,000 to 100,000 times more sensitive than our own. Where we have about six million olfactory receptors in our noses, some breeds have something nearer 300 million. The part of the brain devoted to smells is proportionally gigantic – about forty times greater than our own.

Science writers have tried many ways to put into context how this compares to our own ability for smell. Dog expert and author Alexandra Horowitz argues it is like us being able to detect a teaspoon of sugar in two Olympic-sized swimming pools. Others argue it is like smelling a single rotting apple in two million barrels of fresh apples. I prefer to imagine the dog's sense of smell in visual terms in a way I can better relate to: to be a dog in a world of smells would be like looking into the night sky and seeing 100,000 twinkling lights, each a distant star – exploding red giants or white dwarfs, asteroids, comets – the lights of time, pitching up and fading out. They would see it all, and they would surely pity the half-experience our senses allow us.

Even the exhalation of the dog is something to behold. As air coming out of the nose is forced sideways between the slits at the tip of the muzzle, this creates a drop in pressure in front of the nose that fresh air (containing new odours) rushes in to fill. This allows a dog, even on an outbreath, to keep the smells rolling in. In one study, a hunting dog in Norway was recorded to have maintained a continuous sniff for forty seconds, spanning thirty inhalations and exhalations.

Dogs can also wiggle their nostrils independently, to make the most of any and each passing smell. Evolutionarily, they've got it all. In the words of Jean Anthelme Brillat-Savarin, the father of food writing: 'The sense of smell, like a faithful counsellor, foretells its character' – the same is true, only more so, in dogs.

It's not that humans have a poor sense of smell. In fact, there are some smells we humans can detect at lower concentrations than dogs. Some plant leaves, for example, are menu items that barely register in a dog's diet and so natural selection has little to ratchet its cogs into. Predictably, it is the meaty smells that dogs are most attuned to picking up. Dogs are highly sensitive, for instance, to the fatty acids given off from dead or dying animals.

It is clear that, in a funny sort of way, dogs can time-travel through smell. Because many odour molecules have a habit of breaking up and falling apart over a short period of time, dogs are able to judge from which direction odour molecules are at their freshest and then make their movements accordingly. Like other mammals, most breeds are quite adept at following the path of a fleeing rodent whose movements are given away by a propensity for urination, for instance.

But there is yet more to their incredible senses than this. As of May 2020, we can add a new sense to the repertoire of dogs. We now know that their noses are able to detect thermal radiation too. Dogs, it appears, can heat-seek. To discover this, researchers from Sweden and Hungary first trained dogs to pick out identical objects with different temperatures. Then, using an fMRI scanner, they looked at the brain function of thirteen different pet dogs when provided with neutral stimuli and with warm thermal stimuli.* When shown warmer stimuli, a tiny patch of neurons in the dogs' left somatosensory cortex activated. This means dogs are in an exclusive club when it comes to extra-sensory endowments – some beetles, certain snakes and vampire bats are capable of the feat too. The extra-sensory talent may be particularly helpful when dogs are digging rodents out of burrows or out from under dense vegetation. It should be noted that this capacity to sense heat pales in significance to that of the nose's primary function: to hoover up odours.

---

* Functional magnetic resonance imaging (fMRI) is a non-invasive technique that measures the small changes in blood flow that take place in specific parts of the brain. It differs from MRI scanning, which makes images of anatomical structures in the body rather than metabolic activity.

With the sense of smell occupying such a large region of the dog's brain, it is obvious that a dog experiences the world in a different way to us – that their perceptions of the world differ from our own. In this sense, Nagel was absolutely correct. Like bats, their world is tuned in to a medium largely unavailable to us. But the question of consciousness remains significantly harder to pull apart. Can a dog know that it is alive? If so, is it conscious of its own existence? Can a dog plan for future events like we can? Can a dog reflect on, lament or feel shame for past actions? Exactly what does a dog's daily experience of being alive *feel* like?

As a new era of cognitive psychology was starting to gather pace in the 1970s, animal scientists and psychologists knew that, in order to answer questions like these and rebuff the ideas of Nagel and others, some sort of test was needed. A test that would somehow expose how aware animals might be of their existence. The test they stumbled upon was staring them, quite literally, in the face.

One of the events said to have shaped Darwin's notion of shared ancestry was his first meeting with Jenny, London Zoo's very first orangutan, in 1838. Acquired by the British Empire (in the worst of circumstances, one imagines), Jenny – or 'Lady Jane', as she was more formally known – was a popular exhibit at the zoo and a young Darwin regularly spent time with her, even entering her enclosure. On one occasion, Darwin brought with him, under his coat, a hand mirror that he passed over to Jenny to gauge her reaction. Her response was electric and not quite what he had expected. Upon seeing her behaviour, Darwin reached for his notebook, and jotted down the events that followed. According to his notes, Jenny was: 'astonished beyond measure at [the] looking glass, looked at it every way, sideways, & with most steady surprise,' he wrote. He continued: 'after some time stuck out lips, like kissing, to glass ... Put body in all kinds of positions when approaching glass to examine it.'

Darwin was mesmerised by this response. He saw great significance in the event. Indeed, it is said to have been something of a seminal moment for his ideas about evolution. At a time when apes were considered by most to be a charismatic, almost comedic perversion of human form, further down the pecking order on the great Chain of Being, Darwin began considering Jenny in almost human terms. After this meeting with her, Darwin wrote that man could no longer 'boast of his proud preeminence' – an attitude that went against the grain of the zoological community at the time, who rather liked the idea of us having our own special place in nature on the moral side of the metal railings, so to speak. But the simple experiment really was seminal. Without knowing it, Darwin had undertaken the first so-called 'mirror test' on a non-human species – a simple test that could show if an animal recognised itself as an individual in the world. It is astonishing to me that no one thought to replay the experiment for another century or so. The person who finally thought to give it another go, in the late 1960s, was the American psychologist Gordon G. Gallup Jnr.

The mirror self-recognition test had simple beginnings. First, working with captive chimpanzees, Gallup arranged for the installation of a mirror and watched how, in those earliest moments, his study chimpanzees responded when seeing themselves. His observations were similar, if a little more stilted, to what Darwin saw with Lady Jane. The chimpanzees showed interest, certainly. At first, their reflections invoked aggressive gestures and vocalisations that were not unlike those you will have seen should your dog or cat be shown a mirror for the first time. But then something more fascinating occurred: the chimpanzees came to understand, apparently, that the figure in the mirror was a representation of themselves.

They began using the mirror to visually groom themselves in places never before seen; to pick their noses or their teeth or inspect or assess their various orifices, taking advantage of the new perspective. The chimpanzees seemed to recognise themselves in a way that other animals (specifically monkeys and gibbons) did not, even after hundreds of hours of trials. From this, Gallup concluded, 'insofar as self-recognition of one's mirror image implies a concept of self, these data would seem to qualify as the first experimental demonstration of a self-concept in a subhuman form.'*

Gallup was convinced this meant the chimpanzees were self-aware – that they looked in the mirror and understood that the animal looking back was them. But some zoologists and psychologists remained sceptical. The object in the mirror could be interpreted as little more than a plaything, they said. Interacting with the figure in the mirror does not suggest the chimpanzee knows it is them, they argued.

To rebut this criticism, Gallup went away and devised a more formal test that could be applied to animals when looking at themselves in a mirror. What he came up with was the mirror self-recognition test – the MSR test. First, the captive chimpanzees in Gallup's care were given ten days to get used to the presence of a mirror. Then, on the eleventh day, each individual was anaesthetized for a short period. While the chimpanzees slept, experimenters daubed with bright red dye a small mark on the eyebrow of each chimpanzee being tested and atop one of their ears. The chimpanzees woke, blurry-eyed, and had no idea what had occurred.

* Over a period of decades, Gallup's research interests have included a host of fascinating research fields, including animal hypnosis, human attraction and the evolution of human sexual behaviour. He is particularly noteworthy for his paper 'Does Semen Have Antidepressant Properties?', for which the answer (following Betteridge's law of headlines) would later turn out to be something of a 'No'.

When each of the chimpanzees was returned to its cage, their response when they first looked in the mirror was observed. If the test subject saw the spot in the mirror and attempted to inspect it, this would suggest they knew the ape in the mirror was them. If they ignored it, it would suggest they did not.

Clearly, in their day-to-day habits, the chimpanzees regularly touched their face and ears. But when the mirror was provided, their touching of the daubed spots went up dramatically – in fact, some individuals picked and pulled at the spots seven or so times when the mirror was present. To Gallup, this was clear evidence to convince the sceptics. It was proof that chimpanzees looked at the mirror and understood that what they were seeing reflected back was themselves. They were self-aware, in other words.

Since Gallup's hitting upon the mirror self-recognition test, other animals have been shown to have a similar talent for self-identification. Passers of the test include bottlenose dolphins, orcas and false killer whales, plus the Eurasian magpie (a non-mammal!) and a single Asian elephant by the name of Happy. Lady Jane's response to Darwin's mirror wasn't a fluke – orangutans (and bonobos) can also pass the mirror test.*

Yet Gallup couldn't quieten all the cynics. One regular criticism, still levelled at Gallup today, is that mirror tests don't always provide the same results within a species. One study involving eleven chimpanzees, for example, saw only three individual chimpanzees respond to the mirror in a way that Gallup described in his landmark study. Another criticism

---

* Interestingly, gorillas perform poorly in mirror tests. This may be down to the fact that gorillas avoid eye contact with one another more than other apes, so the test biases its own results. One gorilla known to pass the test was Koko, the captive gorilla trained in American Sign Language. It is said that, on one occasion, Koko looked in the mirror and was asked what she saw. She signed, in response, 'Me, Koko.'

is that the behaviours may not mean what we think they mean, as Morgan had attested all those years before. Chimpanzees, for instance, regularly engage in long bouts of self-grooming – this is a social behaviour, done alongside others. It may be that chimpanzees engage with mirrors on a social basis, argue these critics, which could explain the response seen in the mirror. There is also the additional problem that some animals that do not quite fit the bill when it comes to cognitive complexity can seemingly pass the test with ease, including cleaner fish, ants and (courtesy of that man Skinner again) laboratory pigeons.

Critics aside, many experts in animal behaviour continue to view the mirror self-recognition test as a useful tool, alongside others, to gauge the capacity through which non-human animals see themselves as players in the world or, to put it another way, the central cast of their own story. Plus, by using stickers that can be subtly glued onto the body, mirror tests can be performed on the cheap, even in one's own home with family pets.

You are likely to have seen for yourself that dogs tend not to do well with mirrors. Upon first interaction with a mirror, puppies may freeze up, bark intently, attempt to play and then run away or look behind the mirror to decipher the strange illusion. What they do not appear to do is take in their face, paw at their whiskers or wink. Interestingly, though, dogs do appear capable of picking up our approach in a mirror or recognising the appearance of us, sneaking up from behind for a tickle. But they just don't treat a mirror like we would. It might be that, cognitively, they cannot comprehend what chimpanzees and orangutans manage. It could equally be argued, however, that they understand what is looking back at them in the mirror and they just don't care. This sounds like a flippant thing to say but it gets to the essence of the argument around the identification of self. In fact, it is another criticism of Gallup's famous test.

Self-reflection means a lot to us. The thing in the mirror reminds us of our place in the world, our standing, our ageing, our timeline, our hopes and dreams – the story of who we are and what we represent. To a dog, one's reflection doesn't need to stand for any one of these things. It is simply light reflecting off a shiny object. For this reason, many have argued – perhaps rightly – that the mirror test saw us shine a magnifying glass on nature only to gaze longingly at our own reflection in its surface.

There are other questions of bias in Gallup's procedure, namely that the mirror test was inherently biased to those animals that see the world visually, like we do, rather than through sounds (like bats) and smells (like dogs). In this way, with some species, the test was flawed. In the same way we might expect Nagel's bats to flunk a test like this, the same is true of dogs. To demonstrate this point, the University of Colorado animal biologist Marc Bekoff undertook his own research on the subject, devising a self-recognition test that was biased towards smell rather than sight. What he came up with (alongside his faithful research companion, Jethro) is delightfully known as the 'yellow snow test'.

The methodology to perform the test is elegant and simple. First, the dog urinates as normal. Then, you, the dog-walker, pick up the slushy urine patch and translocate it to somewhere else on the dog walk. You place it carefully back on the floor somewhere else and then you watch – you observe what happens when the dog finds their own urine in a new place. If the dog is an unseeing, unthinking robot, they should decide to sniff the urine as if it were a conspecific or a competitor, covering the urine with their own splash. Instead, in Bekoff's case, the dog gave their own translocated urine barely a cursory sniff. 'No point in wasting time on one's own urine, after all,' the dog seemed to say. 'On to the next one …'

Bekoff's sample size was small – it really was just one dog – but, as a proof of concept, it went some way to poking holes in the one-size-fits-all approach of the mirror self-recognition test, challenging the test's inherent biases when applied to other animals.

In 2015, Bekoff's approach was built upon by the Italian evolutionary biologist, Roberto Cazzolla Gatti in his paper, 'Self-consciousness: beyond the looking-glass and what dogs found there'. Gatti collected samples from four dogs over four seasons, labelling and storing their deposits and then drawing upon them in a series of experimental set-ups – this was his newly devised Sniff Test of Self-Recognition (STSR). In the STSR, dogs are released into a fenced pen, within which have been placed various urine samples from other dogs, alongside their own and an odourless control sample. The dogs' movements and behaviours in the pen are then observed. Predictably, as Bekoff's simple study had suggested, the dogs showed much less interest in their own scents and cared much more about the scent of the other dogs. The results also seemed to suggest an increased capacity for self-awareness as they progressed into adulthood – a pattern similarly observed in chimpanzees and in humans.

'We would never expect that a mole or a bat can recognize themselves in a mirror,' argued Gatti, 'but now we have strong empirical evidence to suggest that if species other than primates are tested on chemical or auditory perception base we could get really unexpected results.'

Other dog researchers have independently come to such a conclusion, trialling their own methods to test Bekoff's original observations. In one, undertaken by Alexandra Horowitz, metal cannisters featuring different concentrations of urine from a range of dogs suggested similar results. Not only did dogs show more interest in the unfamiliar scents of other dogs' urine, they also seemed to recognise when their own odour had been tinkered with.

Clearly, tests like these suggest that dogs – at least when it comes to their urine – have some understanding of self. They recognise some scents as their own, perhaps in the same way that you would recognise your reflection in a shop window. But how much baggage do dogs bring to these moments? Can dogs ask questions about their self in those fleeting glances? Does the presence of their urine in a place they did not expect to find it bring out questions in their mind? Bluntly, can dogs think about their place in the world? Can they *think* about *thinking*? These are, as Nagel well understood, incredibly difficult questions to answer. In fact, most modern-day animal behaviourists would still baulk at responding to them with any confidence. But, to some of Nagel's opponents in the 1970s, questions like these really did seem answerable.

By the late 1970s, the question of animal consciousness had become something of a hot topic. There were a dizzying number of ideas about the phenomenon to choose from. So much so that a number of new journals began to pop up to cater for their needs, with names like *Consciousness and Cognition*, *Frontiers in Consciousness Research* and *Psyche*. Conferences about consciousness began to appear soon after, where intense feuds between rival camps would play out like theatre. 'The evolution of the capacity to simulate seems to have culminated in subjective consciousness. Why this should have happened is, to me, the most profound mystery facing modern biology,' wrote Richard Dawkins in 1976's *Selfish Gene*.

Not only were these arguments taking place in academic circles, they were also beginning to play out in the public sphere too. The decade saw a host of films and books devoted to the subject, picking up on a irrepressible appetite to understand the incredible phenomenon of human consciousness. The concepts were vast, the themes common. They included robots that had aspirations to be human in *Westworld* (1973), buoyed along by Philip K. Dick's 1968 classic *Do Androids Dream of Electric Sheep?* (which reached cinema as *Bladerunner* in 1982). There was *Solaris* (1972), about a planet that brings out the repressed memories and obsessions of those astronauts who stray too close. The original *Planet of the Apes* franchise (1969–73) reframed our relationship to other primates and saw them, rather than us, play the role of civilising overlords.

Eventually Donald Griffin (he of echolocation fame) would wade back into the debate, giving something of a hard stare to Nagel's assertion that we could never truly know the minds of other species or, indeed, whether they think about things in the same way that humans do. In 1976, Griffin published one of his great works, *The Question of Animal Awareness*, in which he formalised his opinion that humans and non-human animals have senses and brains that differ from one another on a continuum rather than by a dividing line. His central belief was that humans weren't the only ones to have a conscious experience of the world. He argued that other life forms on Earth – including dogs and even insects – had the potential to be knowing actors on life's stage, and that scientists just needed to ask better questions to prove it. The book was contentious, controversial and met with backlash from some philosophers, psychologists and the now-dwindling behaviourists.

'Griffin's subjectivist position, and the suggestion that even insects such as honeybees are conscious, seemed to many scientists to represent a lamentable return to the anthropomorphic over-interpretation of anecdotes seen in Darwin and Romanes,' recalled one account. Many scientists remained unconvinced of Griffin's rather outspoken views. There just wasn't enough science but too many stories and nothing concrete. In fact, so left-field did some of his ideas appear that colleagues considered whether the 'grand old man was slipping into senility'.

'Little data existed to support directly Griffin's thesis that animals engaged in meaningful mental activity – primarily because researchers had been trained to ignore the so-called "anecdotes" that did not fit into the standard paradigm, which rejected anything to do with mind, desire, purpose, awareness, thinking, and consciousness,' scientist Irene Pepperberg wrote about her mentor. Yet, inspired in those heady days by Griffin, Pepperberg would become famous for showing the world the cognitive prowess that parrots possess, blurring and reframing continuing debate about human haves and have-nots. In fact, in a nod to Griffin's influence, Pepperberg named one of her research parrots partly in his honour.* As well as using human language that rivals that of young children, Griffin the parrot possesses a knack for memory games that has seen him beat scores of Harvard students, delighting everyone except presumably the students. Almost certainly, Griffin would have been honoured to have had such a parrot take his name. And undoubtedly, he would be proud to know his former mentee was continuing the legacy he helped begin.

Inspired by Griffin and his bats two decades previously, the 1970s had seen Nagel post his famous question. Like the bat he memorably considered, it flitted around a scientific community, diving, weaving, crossing fields of view and between the ears of the academic establishment. Too fast and too beautifully constructed, like all good philosophical arguments, for scientific inference or explanation.

In time, cognitive psychologists and ethologists would develop the tools required to argue the case for non-human animals more firmly – to see and understand and show others

---

* Irene Pepperberg was kind enough to inform me that 'Grif is named partially after Don Griffin (who was one of my mentors), partially because when we got him as a chick (at 7.5 weeks) he looked a bit like the mythical Gryphon (all beak and talons), and partially because at the time the students in my lab were reading a book series, "Griffin and Sabine" in which a stamp having the image of a parrot plays a role in their love story.'

great feats of cognition in non-human animals. But that would take time and, to a degree, the right study animals. Being so closely related to humans, apes seemed the obvious choice for the cognitive research that would follow in the decades to come. But by looking across at the tender twigs of our recent shared ancestry, we forgot to look down at the animals beneath our feet, that share our lives, that have adapted to our whims, that have, literally and figuratively, grown up alongside us. The animals whose boughs most closely align with the human story.

The dogs, eventually, would have their day again. But not yet.

CHAPTER 10

Flip, the switch

> 'The spirit of speculation is the same as the spirit of science, namely... a desire to know the causes of things.'
>
> – George Romanes (1848–1894)

'A senior colleague took me out to lunch and said, yes, he had the utmost respect for Francis, but Francis was a Nobel laureate and a half-god and he could do whatever he wanted, whereas I didn't have tenure yet, so I should be incredibly careful.' The Francis referred to in this quotation is none other than Francis Crick, one of the discoverers of the elixir-like molecular helix we know as DNA. The man recalling this conversation? Christof Koch, who was in 1990 a plucky neuroscientist with whom Crick had been working on the problem of consciousness. 'Stick to more mainstream science!' Koch remembers being told during this encounter. 'These fringey things – why not leave them until retirement, when you're coming close to death, and you can worry about the soul and stuff like that?'*

We like to think scientific questioning has no limits. We like to think of scientists gazing upon everything – every single natural phenomenon they encounter – with a questioning gaze and a thirst for answers. But this is not the case. Often science is not like that at all, for the small matter of reputation imposes boundaries upon the questioning minds of many scientists.

To many animal scientists in the late 1980s, consciousness became the wrong kind of scientific question. It became sullied – impossible to grapple with, contentious and controversial. Some biologists began to consider animal consciousness as something of a backwater. A place where only the scientists with least to lose, like Francis Crick, were confident to tread. Things got so bad that, in 1989, the British psychologist Stuart Sutherland would declare of consciousness that 'it is impossible to specify what it is, what it does, or why it evolved. Nothing worth reading has been written on it.'

During this decade, the study of dogs and their cognition had drifted into a similarly defunct backwater. Where once dogs had garnered the attention of geneticists, behaviourists and evolutionary biologists, now the use of dogs in cognitive research fell into a steep decline. Most zoologists steered well clear of them, for a variety of reasons. First, the comparative approach that zoologists normally apply to

* Needless to say, Koch did not listen. He went on to author more than 300 papers and five books on consciousness and currently sits as president and chief scientist of the Allen Institute for Brain Science in Seattle.

animals when measuring differences between closely related species and speculating on their evolution, for instance, just didn't work as well with dogs. After all, which adaptation did we endow upon them and which adaptation did nature? It was all a bit ... hard to place. Plus, there were methodological problems. Working with pet dogs introduced hard-to-control variables that weren't a problem when studying wild animals, such as how a dog might have been treated, what he or she had been fed, or how socialised a given dog might be. That problem didn't exist with lemurs or slow loris or lions or lion tamarins. To understand the evolution of animal forms, zoologists argued, you had to go to the source: nature, the outdoors, the wild. On this scientific frontier, dogs just couldn't deliver. Plus, there was the simple issue of physiology: dogs (like many domesticated species) have a brain 25 per cent smaller than their wild cousin, the grey wolf. What scientist would want to work with that? To many, it was like doing a jigsaw with only three-quarters of the pieces.* 'Back in the 1980s, the study of any kind of domestic animal, let alone the ultimate artificialities that are pet dogs and cats, was deeply unfashionable,' writes John Bradshaw in *The Animals Among Us*. The cognitive ethologist Mark Bekoff, who made a career studying dogs, recalls being asked early in his career: 'Why don't you study *real* animals?'

There may have been other reasons that dogs had lost favour with the animal cognition research community. To a degree, fingers had been burned in earlier decades by proclaiming dogs as wolves of the home, with masculine

* Quite why dogs have smaller brains than wolves is still a matter of debate. Interestingly, it is an observation recorded in other domesticated animals, including horses, pigs and ducks. You and I are not immune to this phenomenon. Human brains are about 10 per cent smaller than they were 10,000 years ago.

(and, as we have seen, mistaken) notions of dominant alphas, looking over at us from the dinner bowl, waiting for the moment to strike and take over the household. Mistakes like these were too easy to make with dogs. The ground felt too uncertain. Besides, Scott and Fuller's studies had proven how damaging the pursuit of scientific knowledge could be on the temperaments of the dogs with (and upon) whom scientists worked.

For these reasons, looking to dogs for a new perspective on the evolution of animals was like looking to KFC for a new perspective on the wild ancestor of chickens. Reputational suicide. And so, rather than look to dogs for answers about animal cognition, many researchers instead maintained a laser-like focus on chimpanzees and bonobos. They were, after all, our closest cousins and so this is where researchers expected to find the greatest feats of animal cognition. In a decade or two this attitude would change, but we weren't there yet.

At the time, as the 1980s were dawning, two of the biggest names in animal cognition were David Premack, animal behaviourist, and Guy Woodruff, psychologist. Just like Gordon Gallup Jnr (made famous for his mirror studies) the pair focused a great deal of their time on captive chimpanzees. They were interested in a sub-element of the animal cognition debate. Their hunch was that chimpanzees could understand and reflect on the intentions of other individuals, and then attempt to use this information to manipulate others to meet their own aims. Woodruff and Premack were keen to explore whether chimpanzees could, in other words, deceive others to get what they wanted.

One of the experimental set-ups favoured by Woodruff and Premack involved something of a hint of good cop, bad cop. The 'good cop' in their test was a kind, co-operative trainer who, when finding hidden foodstuffs in an enclosure, always gifted some of it to a watching chimpanzee. The 'bad cop' was a mean, uncooperative trainer who never shared edible treats with the chimpanzees when they were pointed out by the captive chimpanzees. If spotted, this 'mean trainer' chose to eat the food reward immediately upon discovering it, gifting none to the watching chimpanzee. The question was: might the chimpanzee cotton on to a strategy to get the food rewards each time? Might they, somehow, choose not to tell the mean trainer about the food and leave the rewards there for when the kind trainer returned? A human could work out the best play, sure, but could one of our nearest and dearest cousins?

The results Woodruff and Premack obtained would prove hard to interpret. Of the four chimpanzees involved in the study, three failed to show evidence of deceptive behaviours. It seemed they didn't opt to deceive – or they failed to display the cognitive skills to manage it. The fourth, however, was more successful. This chimpanzee apparently chose not to tell the mean trainer about the food. Eventually, it really did appear to catch on: the chimpanzee was seemingly capable of using deception to get results. The results were encouraging, but Woodruff and Premack needed more.

A few months later, the pair tried again. This time, two of the chimps did keep the location of the food reward a secret from the mean trainer. And, interestingly, they went one step further: they actively pointed to the wrong location of where the food reward was. Quite simply, these individuals told a bare-faced lie. Just like humans, the researchers deduced, the chimpanzees were capable of deception. They could imagine the viewpoints of others and change their behaviour accordingly. They were demonstrating what Premack and Woodruff called a 'theory of mind' – they could imagine seeing the world from a perspective that differed from their own and change their behaviours accordingly.

At the time, the chimpanzee study was widely hailed as something of a breakthrough. It became one of the benchmark studies in how the cognitive divide between humans and chimpanzees was smaller than the 'old guard' of science had anticipated. But the truth is that lying did not come easily to these chimps. In fact, it took several hundred tests (over many months) before the deception behaviours began to kick in. Even then, one chimpanzee never quite got it.

Predictably, it wasn't long before Woodruff and Premack's critics came out of the woodwork. They argued that the results could easily be explained away as differential reinforcement – that the study animals had learned, through operant conditioning, to repeat behaviours that led to food rewards. From Woodruff and Premack's supporters came a different argument: the experiment had taken a while to deliver results because the chimpanzees had problems understanding the parameters of the task. Debate raged.

What chimpanzees know about the minds of others, and how they can act on this information, remained something of an impenetrable mystery. And so, in the mid-1980s, psychologists turned to another study animal for answers: they turned to human children to establish how and when their theory of mind develops. The psychologists brought two famous puppets, called Sally and Anne, with them and devised a test to assess the cognitive abilities of the children involved in their studies.

In the Sally-Anne test – a simple routine that measures an individual's ability to attribute false beliefs to others – a child is shown a little theatrical play that features a cast of two puppets and props that might include a bed, a box and a ball or a marble. First, puppet Sally plays with the ball before putting it into a box and leaving the room. While Sally is out of the room, puppet Anne enters the room and takes Sally's ball out of the box and puts it under the bed. Anne then leaves the room and Sally comes back in, eager to get her hands on the ball again. At this point, it's time for the big question. The psychologist turns to the child and asks: 'Where will Sally look for the ball?'

The insight this simple test can give into a child's developmental state is truly fascinating. Three-year-old children mostly answer the question with an almost incredulous tone: 'Sally will look under the bed,' they'll say. Their confidence comes from the personal experience of what they have just watched – 'I just saw it, so that's where she'll look!' Four-year-olds, on the whole, know better. 'Sally will look in the box,' they'll say, quite correctly. This is because they can imagine what it is like to be Sally. They can devise a theory of another mind. This, psychologists argue, is a key developmental stage in the human understanding of the world. At this point, young children begin to see their mind as an entity that is one of many and to understand that other people carry in them perspectives that differ from their own. They become able to theorise about the minds of individuals other than themselves.

Naturally, animal behaviour scientists began to wonder if such a test could be devised for their study animals. While dogs continued to labour in the shadows, apes were once again selected as the most suitable audience for such an experiment. Some scientists stuck close to the Sally-Anne principle, using characters moving things around a scene while captive apes watched, anticipating what might happen next. But, again, it was hard to draw firm conclusions from the results in studies like these. Just as with deception studies, they provided interesting snapshots – something was going on in their minds – but proving beyond belief that it was true 'theory of mind' in these animals was still too far a stretch. Many agreed it was there somewhere – indeed, numerous observations of wild apes that were taking place at the time seemed to suggest it almost certainly was – but theory of mind was hard to pull out crisply from the complex miasma of everyday behaviours that apes exhibit.*

In many ways, the same criticisms that had been thrown at Gallup's mirror test studies were thrown at those who supported the idea that animals other than humans had a theory of mind. Just as an animal failing to gaze longingly at itself in a mirror is not proof that it understands its reflection, an animal failing to respond like we would to a human-centric narrative drama does not mean they lack minds like our own. In fact, if apes do have a theory of mind like ours, there is very little reason to assume it will be energised by a fairly mundane story of a girl who hides marbles under a bed.

In time, though, some answers about theory of mind in animals would come. And when they finally did arrive, the cognitive gift would expose itself in an entirely different way, from a different branch of the mammal family tree entirely. Not ape, nor monkey. Not dolphins or whales. It was dogs who provided the much-needed perspective and gave us an idea of what theory of mind might look like in non-human animals. The story was moved along by a dog called Flip. A dog who helped build a research institute that would change forever the way we study the minds, behaviours and cognitive talents of humanity's best friend.

Flip doesn't look like a typical hero – he has no cape, no obvious swagger, no status. He's a bit fuzzy, low in stature,

* The debate about whether and to what degree apes have theory of mind continues to rumble on in the current day. In 2016, Fumihiro Kano and colleagues from Kyoto University (using a degree of ape cosplay) measured gorillas, chimpanzees and bonobos looking at objects in a Sally-Anne-like scene to argue that they look longer at the location where the returning character should look.

mixed breed – an everyman kind of dog. In truth, Flip has very few cinematic qualities. His origin story is notable, however. Like Superman, Flip appears to have almost fallen from the sky on that fateful day in November 1989. And, like Superman, he is taken in by a kind family who raise him as their own and see in him something no else does: potential. There is no Smallville in this universe. The first meeting with Flip occurs in Hungary's Kékes mountain, where Flip is found roaming a cafeteria, clearly lost and somewhat out of his depth, by Vilmos Csányi, a scientist who was then head of the Department of Ethology at Eötvös Loránd University in Budapest.

In the moments that followed this meeting, Flip took to Csányi and Csányi took to Flip. So much so that, within hours of their meeting, Flip became part of the furniture of Csányi's life. He moved in and never left. In the weeks and months that followed, Csányi – an ethologist through and through – began to observe interesting things about Flip's behaviour. How quickly Flip was able to tune in to the routines of his human companions, for instance. Walks, car journeys, food – in each case, Flip would always seem to know what was coming next. This interested Csányi a great deal. Flip became something of a professional interest.

Like all good scientists, Csányi began keeping a diary about Flip's behaviours, with the entries recounted throughout his fascinating book, *If Dogs Could Talk*. The entries are collections of moments, really. Upon seeing Csányi pack for a holiday, for example, Flip begins to mope around and so Csányi shows him his walking stick and tells him that it's an excursion Flip will be joining them on – Flip goes wild with joy at this. In another diary entry, apparently with intention in mind, Flip finds Csányi and guides him to a closed door behind which a dog playmate has got stuck. On another page, we hear of the time Flip walked up to Csányi with purpose, attempting to inform him that there was something he needed help with. Flip did need help: his water bowl needed refilling. Within weeks, there is a diary entry where a rain-soaked Flip purposefully runs inside and rubs himself on the mat: 'Do you want a towel?' asks Csányi. With great delight, Flip runs upstairs to where the towels are kept. There are many others like this.*

Throughout *If Dogs Could Talk*, Csányi is quick to remind readers that these observations are anecdotes, not data. But diary entries like these became a way for Csányi to consider bigger ethological questions about what goes on in the minds of animals and how we might test for them. 'I know that one dog or story does not prove anything,' he writes, 'but a faithful record of a recurrent or peculiar dog story will surely help in the design of new experiments and theories.'

Csányi was never Flip's master. It was the other way around. It was as if Flip had shown an uncanny knack for mastering Csányi – learning his routines, his ways, his tells. 'Dogs are excellent human ethologists,' Csányi notes (to which the obvious riposte is: takes one to know one). Csányi was eager for his colleagues at the Department of Ethology to take seriously the idea of using dogs to answer big research questions about animal minds. Yet, at first, not everyone shared his newfound excitement.

'Csányi is very good at storytelling,' recalls Ádám Miklósi, editor of *Dog: A Natural History* and now head of the Ethology Department at Eötvös Loránd University, 'but we had no idea about how we could "transform" these stories into meaningful (simple) experiments.' In those early days, this was clearly a challenge. 'First, there was at that time no

---

* Among my favourite of the entries in Csányi's book is the moment where Flip, apparently on his best behaviour, appears to ask Csányi to help dig up a molehill, by looking upward questioningly and gently tapping the mound with his paws. Sternly, Csányi tells him 'No.' 'He gave a big sigh and lay down next to it,' Csányi writes.

research on dogs,' Miklósi explains. 'So we had no idea where to start. It was also not trivial to work with family dogs. And the ideas on how dogs may learn words, understand human communication, seemed a bit ... far-fetched.'

In time, Csányi's enthusiasm won out. Five years later, the department would change to direct its research efforts far more extensively upon dogs. The research team would focus on their cognition, their evolutionary history – on dog–human interaction, a broad term that encompasses their communication, co-operation, their social learning and how they follow rules.

Together, Csányi, Ádám Miklósi and Hungarian ethologists including József Topál and Antal Dóka came up with a plan. They would open a new institute to learn about dog cognition and explore in more depth the human–dog relationship. In 1994, they launched 'The Family Dog Project' and, with it, a new era in dog cognition studies was born. In our correspondence during research for this chapter, Miklósi is a little bashful about the legacy the project has had on the scientific research community ('we may deserve a line at least in the future textbooks,' he writes). The truth is, in scientific terms, the human–dog relationship would look very different without their diligent, hard graft in those early years.

One of the notions that Csányi, Miklósi and colleagues went on to challenge was the popular idea that dogs were (to coin a phrase used by some researchers) 'dumb wolves'. In that aim, one of their earliest research papers would challenge a pre-existing study that suggested wolves were smarter than dogs because, when shown by a human how to open a latch on a gate to get food, wolves happily took to solving the challenge themselves in a way that dogs did not. The Hungarian research team set to work, gently teasing apart the assumptions made in the study and, in this, they succeeded. They found that the dogs involved in the study *could* have solved the gate-latch task in the same way the wolves had; it's just that they were waiting for their human companions to help – they were well trained, polite and well behaved, in other words. It wasn't that wolves were smarter than dogs. It was because the dogs *expected* to work with their owners. The team published their results in 1997.

Another important study came the year after this. Miklósi and colleagues confirmed that dogs that are highly bonded with their owners exhibit a form of separation distress when isolated that is directly comparable to that seen between mothers and infants.* Infants and puppies show the same tell-tale joy and excitement at being reunited, often with frantic desire to touch and embrace – the same energy, the desperateness of it, even down to bounding up and down in a bid for closeness. In the words of psychology, the research team confirmed that animals other than humans can exhibit with one another a social attachment – a psychological term invented in the 1970s to describe the incredibly strong bond that exists between mother and infant. Through this research, dogs became part of an exclusive club.

In recent years, the insights that have come out from studies undertaken by the Family Dog Project are nothing short of staggering. In all, its researchers have produced more than a hundred papers to date, with many more in the pipeline. Among other things, their research has helped us demonstrate and understand great swathes of dog behaviour: that puppies naturally appear to prefer human company to other dogs, for instance; that dogs can be trained to imitate human actions; that dogs can understand that some rules are rigid and some are flexible even though humans are often inconsistent in their rule-making; that, even if their eyes are telling them a

* This conclusion confirmed what had been suggested in 1952 by the esteemed (and largely forgotten) dog trainer Ramona Albert. Albert was the first to articulate distress-anxiety theory of destructive behaviour that occurs when dogs are left for periods by their owners.

ball has been placed under a given cup, the presence of a human telling a dog that the ball is under a different cup is where their attention is better placed. Yet these incredible studies are just the tip of the iceberg. In fact, it is almost as if their understanding of dogs (and ours with it) is increasing at something approaching an exponential rate.

For those who consider this mere hubris, here is a snapshot of the discoveries that members of the institution have contributed to in recent months: that the run-up to dog bedtime affects the quality and type of sleep they have; that their sleep contributes to memory consolidation in the same way that our own brain works; that assistance and therapy dogs are better problem-solvers than other family dogs; that dogs are naturally drawn to patterns of dots that move in a way that suggests human movement; that some dogs are able to categorise toys like humans; that dogs can behave as if they have an episodic-like memory; that dog personalities change as they age; that old dogs show less social interest, struggle with memory tests and are less interested in novel items; that companion pigs are less likely than dogs to look for human help on unsolvable tasks; that assertive, older dogs may be perceived as being the most dominant dogs in multi-dog households. And that is, as I say, a snapshot.*

But this chapter is about the early days of the Family Dog Project – when dogs were still new to the cognitive research arena and their return to the sciences was just beginning. After the publishing of the social attachment study in 1998 came another important discovery. It was one made by Ádám Miklósi and colleagues and simultaneously hit upon by dog scientists working in the USA. For those scientists interested in animal cognition, and with an inkling that the truth of

* The full list of publications (which is updated regularly) is available on the Family Dog Project's website: http://etologia.elte.hu/hu/publikaciok/.

animal minds lay in apes, the results of this study were something of a game-changer. Dogs, it seemed, had a trick up their figurative sleeves. They were capable of doing something that apes could not. They could read our gestures and respond accordingly. They could apparently – in a kind of Rubicon moment – interpret the significance of that most sacred human gesture: the pointed human finger. This single discovery made many parts of the animal cognition community sit up and (slowly) take notice.

The American side of the 'pointing' story is worth telling, partly because it has become a wonderful demonstration of how good science progresses through observation, through experiments, through interpretation and through a respectful amount of heckling from the sidelines. The story begins with the comparative psychologist and linguist, Michael Tomasello. In 1994, Tomasello was fast becoming one of the great thinkers of our time, particularly on the subject of animal cognition and the similarities and differences between humans and non-human apes. His lectures at Emory University, where he was based at the time, were well-attended affairs – the audience, naturally, were very keen to hear the perspective he had acquired from many years of studying the cognitive capacity and wherewithal of apes, including humans.

At the time, his research focused on the evolutionary reasons behind the human rise to power. *Why were we humans reading books and building rockets and not, say, chimpanzees?* he wondered. Key to Tomasello's hypothesis was that humans differed from other apes in a number of defined ways. One of these differences, he argued, was that humans can view events from multiple perspectives (think: 'theory of mind') in a way unmatched by other primates. Another, he reasoned, was that humans, unlike other apes, can imitate the actions of others in a nuanced and carefully considered way. He also argued that humans differ from other apes in our capacity to build 'cumulative culture' – to pass on, share and improve upon our discoveries, generation by generation. Finally, Tomasello also asserted another key difference: that humans were the only animals on Earth to truly understand informative hand gestures – that we can both point and get the point, so to speak.

This particular assertion of Tomasello's has become the stuff of legend to anyone with an interest in the history of animal science because, during one discussion in which he was outlining his ideas to students about the human uniqueness of gestural communication (based on fifteen years of researching animals and publishing papers, don't forget), a student politely uttered words that would shape cognition science for decades: 'My dog does that,' came the student's profound blurt. At this, the story goes, Tomasello paused. He considered how best to respond. Then, turning to the student who'd had the gall to interrupt, he voiced the words all good scientists should frequently wield: 'Oh yeah? Prove it!' And so the young student by the name of Brian Hare did just this. Days later, Hare would triumphantly walk back into university holding the evidence – a VHS cassette of his retriever Oreo playing fetch, guided by Hare's outstretched finger.

The video footage showed Hare first throwing a ball, which Oreo set off to pursue. Then, while Oreo's attention was taken with the first ball, the footage shows him throwing another ball in a different direction. Once Oreo returned with the first ball, Hare pointed to the other ball and Oreo hurried off to collect that ball too, diligently and with apparent joy following the direction of Hare's outstretched pointing finger. Tomasello watched the video and was duly impressed. Then (in another example of healthy science in action) Tomasello conceded. To that end, together, they agreed to find out more.

'Oreo led us to the genius of dogs,' Hare explains. 'Dogs might not be able to untangle themselves from a lamp post, or know not to drink out of a toilet, but they are smart where it counts. They can read and understand our gestures in a way no other animals can – not even our closest living relatives, chimpanzees and bonobos.' In the modern day, Hare (now a professor at Duke University and something of a doyen of cognitive science) still draws on his early experience with Tomasello: 'Instead of being offended that a 19-year-old had contradicted Mike's signature theory, Mike was excited – he didn't care about being right, he wanted to know the truth,' he tells me. 'I always give this example to my students to illustrate what a great scientist is.'

Tomasello stayed true to his word. In the months that followed, the two set off, with colleagues, to work on devising a proper scientific test to show that Oreo, and perhaps other dogs, could understand gestures such as pointing. In the experiments, Hare and colleagues used two cups and hid food rewards under only one. Oreo was brought in and Hare would point at a cup, which Oreo would consistently approach, whether or not food was present. Hare brought in other dogs (including a six-month-old) to test their skills and, on the whole, these dogs were also able to complete the task, obtaining scores better than chance.

Interestingly, Hare and his colleagues later tested the experiment on wolves. Wolves, it appeared, struggled with the task. As did chimpanzees. Dogs were clearly able to attend to human gestures in a way that other animals could not, they deduced. Hare's working hypothesis in the late 1990s was that, through the process of domestication, dogs had found themselves selected for a suite of socio-cognitive traits that allowed them to communicate with humans. In other words, natural selection favoured dogs most competent at reading the signals afforded them by humans. Pointing, it seemed, was one of those signals.

This was the same conclusion that Ádám Miklósi and colleagues at the Family Dog Project had reached at about the same time, based on a research set-up that was broadly similar to Hare's. The two research papers – Hare's and Miklósi's – were published within months of one another and were both well received. But it would still take another ten years for the significance of these results to bed in.

'The response was quite gradual,' Miklósi tells me. An article in *New Scientist* magazine in 1999 (titled 'Almost Human') got things rolling and the paper on social attachment took about ten years to gain traction. Then came an international conference, the 'Canine Science Forum' in Budapest in 2008. 'I think that really helped a lot,' Miklósi explains. 'And then, in 2009, there was also a review in *Science* on dog research ... so slowly dogs became the main species in animal cognition by 2010.'

The Family Dog Project is one reason why dogs became such a big deal in the cognitive research community. Its approach, in terms of its ethics, its cost and its results, caught the eye of many other academics and – partly because of the jaw-dropping science coming out of the institution – a wave of dog cognition centres around the world subsequently sprung up. Today, Portsmouth (UK) has the Dog Cognition Centre. At the University of Western Ontario (Canada), there's the Dog Cognition Lab. In the US, there's the Canine Cognition Centre (Yale), the Duke Canine Cognition Centre (headed up by Brian Hare), and the Dog Cognition Lab (Barnard College, Columbia University). In Austria, there's the University of Vienna's 'Clever Dog Lab'. In Australia, there's La Trobe University's Anthrozoology Research Group. In Japan, there's the Companion Animal Research at Azabu University. In Italy, there's the Canis Sapiens Lab. Still, Miklósi remains grounded about the Family Dog Project's contribution to science.

'I hope to be remembered by our ethological approach that we were interested less in "pure cognition" but in the nature of human–dog interaction and how dog evolution led to human directed social competence,' he tells me. 'We have introduced a lot of interesting ideas, like attachment, the study of barking, social learning and imitation, word learning. I think all these different approaches and projects have had a cumulative effect on dog research.'

Many of these research institutes still work on the same principle that the Family Dog Project began with: that the dogs used to support their research are free to come and go; that the dogs have the same rights as humans who partake in psychological research; that the owner–dog relationship comes first, always. If you want to imagine what this looks like, look no further than American journalist Colin Woodard's observation, after visiting the Hungarian set-up: 'Canines have the run of the place, greeting visitors in the hall, checking up on faculty members in their offices, or cavorting with one another in classrooms overlooking the Danube River, six floors below.'

In this vein, without a strap in sight, the 1990s saw dogs re-emerge as an important member of the science fraternity once more; they became research subjects again, on something closer to their own terms. They were ethological subjects, like sticklebacks and jackdaws; animals being studied in their natural environment, which was in the company of humans. From cadavers to collaborators in little more than a hundred years, Flip and Oreo were something like Darwin's dogs – thinking companions, learning allies. Always friends.

As the years ticked on, dogs and people had reconvened at an exciting crossroads in cognitive science. From hereon in, they would move forwards together, in step with one another. Dogs and their minds would no longer be fringe. Far from it. Dogs were becoming mainstream again, and biological science would be made immeasurably better as a result.

CHAPTER II

The power of play

> 'PLAY is older than culture, for culture, however inadequately defined, always presupposes human society, and animals have not waited for man to teach them their playing.'
>
> – Johan Huizinga (1872–1945)

Where most dog scientists look back in time to find clues about how dogs found their way into our lives, Dmitry Belyaev looked to the future. He set out to create dogs again, from scratch. From foxes.

One of Russia's foremost geneticists, Belyaev looked to the silver fox *Vulpes vulpes* – a common species used in the Russian fur trade – and saw an opportunity to begin his own breeding programme. Starting in the 1950s, he would, generation by generation, breed from the tamest foxes in each generation and see how long it took to domesticate them. What Belyaev created is one of the longest and most enlightening experiments in genetics ever undertaken. Not that many people knew this in the 1980s, of course. The project was four decades in before it would become better known to the Western world, courtesy of a 1999 article in *American Scientist* released a decade after Belyaev's death and written by his protégée Lyudmila Trut. (Almost ninety years of age at the time of writing, Trut continues to run the project to this day.)

Belyaev started with 130 wild-type silver foxes and selected those he deemed the most tame. He walked the cages with food in his hand, seeing which individuals would approach and which would bite, which foxes growled and which hid in their beds. Of those that approached, some allowed themselves to be touched and patted. Some foxes even whimpered, clearly warmed by the experimenter's touch. Belyaev was most interested in these individuals. These would be the foxes bred in the next generation. By breeding for tameness in subsequent generations again and again, Belyaev would come to shape their descendants. To all intents and purposes, he domesticated a new sub-species of fox.

By the time of his death in 1985, the silver foxes Belyaev had bred were very different in temperament to those he started with. They were positively excited by human company. Instead of snarling with their teeth, they smiled with their eyes. They wagged their tails with apparent glee. By the end of the century, three-quarters of the foxes were in the 'domesticated elite' category – they actively sought out human company. They whimpered, they sniffed, they licked. Like dogs then, but not quite. 'The amazing thing was that cubs who had just started to crawl, opened their eyes and started showing affection towards humans,' Lyudmila Trut told the BBC in 2010. 'This kind of response was a big surprise to us.'

Particularly interesting in Trut and Belyaev's research was that, by selecting for tameness in their genetic experiments, the pair saw their captive foxes change physically over the generations. Just as with dogs, a proportion of the once-silver foxes were now born with a black-and-white (piebald) coat. Their ears were floppy like those of dogs and their tails curved around over the body in an occasionally ungainly manner. They had (to use a non-scientific term) pudgier faces too – shorter snouts and wider jaws.* These genetic characteristics had clearly hitched a ride with Belyaev's selection for tameness, exposing themselves bit by bit, generation by generation. In all, across half a century, Belyaev and Trut's tame foxes ended up differing from their wild counterparts across forty genes – quite the feat.

Belyaev and Trut's domesticated foxes have become something of a celebrity species to geneticists. They regularly appear on TV science documentaries about evolution, about artificial selection (the process coined by Darwin in which humans, rather than nature, are the selecting agent), and genetics, particularly how traits (such as the piebald coat) often piggy-back onto the same pieces of genetic code, like

* A strangely similar, wholly unplanned, domestication of wild red foxes may be happening in parts of Britain. In 2020, Kevin Parsons, an evolutionary biologist at the University of Glasgow, compared measurements of historical collections of fox skulls, both from urban and countryside settings, collected from culling practices undertaken between 1971 and 1973. One key finding was that foxes living in urban environments had noticeably shorter and wider muzzles and smaller brains than their rural fox counterparts – an observation consistent with that seen in dogs and these silver fox experiments. The hypothesis is that the foxes are adapting to a new food source (garbage) around human habitations and in so doing are domesticating themselves. This might be a useful model to imagine the early human–dog relationship.

those that code for brain hormones. The domesticated silver foxes can also follow pointing gestures, almost as well as dogs.

But there's something else they do too. Look closely at these gorgeous foxes and it becomes clear: these domesticated silver foxes have an insatiable appetite for fun. In video footage, they can be seen to tug. They tussle. They frequently engage in bouts of rough-and-tumble. Throughout much of their lives, even into adulthood, the silver foxes in Belyaev and Trut's experiments are drawn to that most mysterious of mammalian behaviours. They play.

Play is so common to mammal life that it is easy to overlook the significance it has in biological terms. There are dolphins that blow bubbles to one another or surf gnarly waves together. There are elephants that make mud-slides, and wallabies that play peek-a-boo in the pouch. There are rats that pull one another's tails, and kittens that sit stealthily in empty cardboard boxes, waiting to strike at siblings. Even bats play. If there is intelligent life on other planets, there is no guarantee we will see playfulness in its inhabitants, more's the pity. Life, broadly speaking, didn't have to be this way.

For such a common behaviour across species, it might surprise you that we know very little about why play occurs. In fact, for years the phenomenon was considered, like consciousness, little more than a zoological niche interest. One only needs to look at the lack of scientific papers on the topic to see how little interest this common mammalian behaviour yielded in times past. 'Study of animal play historically has languished at the fringes of behavior research,' wrote one animal researcher in 1981. 'Suspicions that play did not really exist or that it was a trivial phenomenon kept many investigators away. However, like an unwelcome guest, play continued to present itself to field workers.'

In those days, play was defined in a fairly nebulous way, which might translate better as: you know it when you see it. According to the animal behaviourist Marc Bekoff, writing in 1984, play encompassed a wide range of natural animal behaviours: the 're-tooling' of facial gestures and body postures to let others know of intentions to mock-fight or wrestle; complex role-reversals or reframings of dominance relationships with others in a group; chasing-fleeing relationships between individuals; self-handicapping behaviours where contact is maintained longer than it otherwise would be with smaller or weaker individuals.*

The functional significance of mammalian play is still widely discussed today. The fact that play seems to peak at puberty – about the time the brain's cerebellum goes into a developmental overdrive – suggests, to some neuroscientists, that play is a way of building and strengthening general connections in the brain, which may enhance the speed and efficiency with which behaviours can be accessed and implemented in adult life. Play may also be adaptive in that it helps animals experience novel situations that they would not otherwise face, devising solutions in a low-stakes situation that they may rely on later in life. The truth is that we don't yet know for sure what play is for. Yet what is clear is that, for an activity that burns up a lot of energy, natural selection would quickly pull it to heel should it be evolutionarily maladaptive. Instead, it is a mammalian universal. Play *matters* to the lives of animals. Unsurprisingly, it matters a great deal to dogs too. Dogs that are denied play-time can become anxious, aggressive and unruly.

As part of researching this book, I have found myself sitting in the park many times, watching dogs of many breeds go about the business of play. The familiar bouncing trot that

---

* Gordon M. Burghardt's 2014 definition is one of the finest that we rely on today: 'Play is repeated, seemingly non-functional behavior differing from more adaptive versions structurally, contextually, or developmentally, and initiated when the animal is in a relaxed, unstimulating, or low stress setting.'

one dog will do when it approaches another, toy held proudly in mouth. The stretching of the forelegs on the ground, hindquarters held high, and the resting of the head on the floor. The wide eyes and raised eyebrows. The panting. The open jaws, smiling. The chasing, the sprinting. The running, turning and rolling over. The exposed tummies and the lolling tongues. Occasionally, there is the glaring of teeth or an assertive woof during these playful interactions. When this happens, most owners stop their conversations with one another and lean in with apparent concern. No need. In all but a handful of cases, the dogs resolve the situation without human help, through tiny gestures or movements so quick that our eyes cannot keep up. Play continues, often with great joy. One dog chases the other, then the other takes its turn, and so on and so forth until exhaustion kicks in – in every park, on most beaches, in fields, across the wilds and in the gardens of millions of dwellings on planet Earth every single day. Play is one of the universal languages of dogs, and they speak it fluently.

Alexandra Horowitz is something of a world expert on the subject. In the past, Horowitz's research interests have seen her study hours and hours of slow-motion video footage of dogs in a bid to understand what this tells us about what the minds of dogs are capable of. She explains many of her discoveries in the fascinating book *Inside of a Dog*, in which the vernacular of dog play is laid plain for all to see. First, in the run-up to play, one will often see the 'play-bow' – the stretched forepaws and head held low. In this position, the rump is high and the tail is wagging; the mouth is open and relaxed. The position is so easy for dogs to gauge that even humans can mimic the position to incite their interest. But play-bows aren't always pronounced and obvious. Some dogs, if they know one another, develop a kind of shorthand to their play-bow – a quick pounce upon the ground like a clap might be all it takes to signal their intentions. (Horowitz calls this a 'play slap'.) Then there's the 'head bow' and the 'open mouth display' – both are established shorthand aimed to elicit the act of play. These signals can do more than just elicit play, however. They can also be employed *during* play as well, such as if one dog goes in a bit too hard with a playful nip and wants to assure the other dog it was all in the name of fun.*

Rolling over is another action that dogs utilise in their repertoire of play behaviours. Though it looks simple, the activity is far more complex than we once imagined. It won't surprise you to hear that, for many years, this behaviour was considered a 'submissive response' – a way of putting the brakes on play or a kind of white flag to an apparent 'alpha'. The truth is actually far more interesting. Careful analysis of dog–dog interactions (including large vs small dogs and vice versa) show that rollovers are far more complicated than just displays of submission. First, dogs that lie on their backs don't remain passive – they may continue to engage in sparring and lunges and faux-attacks at the neck of their playmates. Second, rather than put an end to play, many rollovers actually help play fights continue – they are a sign to keep the fun going. But there is a third interesting thing about rolling over: in some cases, the action can be a way for dogs to 'self-handicap' themselves so that their play can be adapted to suit different sizes or temperaments of other dogs. In effect, by rolling over, a larger dog can give a smaller dog a fighting chance. (Horowitz calls this a 'self-takedown'.)

Play can also influence the relationship that dogs can have with one another and with us. When tested before and after play bouts with humans, it appears that, after a bit of playful interaction, dogs become subsequently more obedient and

* Humans employ a similar strategy – a jovial pulling of the leg is rarely delivered without a broad smile or a cheeky laugh, our own version of a play-bow. Indeed, chimpanzees regularly employ a 'play-face' to get rough-and-tumble started.

attentive to our requests. This supports the common belief that training dogs after a play session gets better results than rewarding them with play-time at the end. It may also be that high-energy games that involve lots of physical contact between dog and human can also help reduce the likelihood of low-level anxiety in dogs.

But play can offer even more than this. In the last two decades, it has become clear that the play behaviour of dogs might provide a novel window into understanding what dogs are able to consider about the minds of others. In other words, through play, scientists are able to understand the degree to which dogs can imagine being the 'Sally' or the 'Anne' in one another's everyday adventures.

Research into this area started to take off in the year 2000, with what became classic experiments in helping us understand dog play behaviours. In these early studies, another renowned dog scientist, Nicola Rooney, (with colleagues from the University of Bristol, UK) recruited a dozen Labrador retrievers and let each roam a large grass paddock with either a fellow dog (with whom it was known to have got along) or a regular human carer. After being given a couple of minutes to scope the paddock out, the experimenters then threw in a prop: a 'tug-toy' – a bit of thick knotted rope with tassels on each tip. No surprises, the dogs quickly became interested in the toy. In fact, within a short space of time, familiar tug-of-war games ensued between individuals in the paddock. What was surprising, however, was how different the games were depending on who the dog was playing with. When playing tug-of-war with a human, for instance, dogs in the experiment would frequently drop the tug-toy, letting the humans 'win' before then actively seeking to restart the game. When playing with other dogs, however, things were far less civilised. In these situations, the two dogs play-fought to gain ownership of the rope and then, once in their possession, they tended to guard it from their canine playmate. Even more fascinating than this was what happened when another toy was thrown into the play arena. If two dogs were present, they would each find a toy and play with it alone, before later coming together again. But with humans and dogs, it was like the second toy wasn't there. The dog did all it could to keep the tugging game going with its human companion. Playing with the person was its primary focus. The dogs were switching play styles depending on the species available. Quite simply, they were making choices.

Other studies, reported by Bradshaw in *In Defence of Dogs*, seem to support these findings. In one particularly memorable study, three actors play out a scene in front of a dog spectator in which a human 'beggar' asks for money from two strangers. One (the 'generous' person) gives the beggar some money; the other (the 'selfish' person) does not. The dog watches this all play out and, once the theatrics are over, the beggar leaves and the dog is allowed to approach either of the human characters. In most cases, the dogs approached the generous person first, choosing to spend more time interacting with them than with the 'selfish' person. The verdict: dogs know if you've been naughty or nice.*

However useful these studies had been, they still only hinted at what dogs might be capable of. To discover more about their cognitive prowess, a different type of test was needed – one in which dogs could interact with another more naturally, in a more natural setting. A play-park, essentially. This is the arena in which Alexandra Horowitz has spent

---

* In their own behaviours, dogs can also 'choose' whether to be naughty or nice. In recent years, studies have shown that dogs change their behaviours according to who is watching. We know, for instance, that many dogs that have been forbidden to take food are more likely to attempt to snaffle it up when a human owner has their back turned or their eyes closed. Likewise, if the human is blindfolded or if the lights are turned down.

much of her time, investing many, many hours painstakingly watching the intricacies of dog play, pulling apart the behavioural two-ways (or three-ways) of play in a manner no one before had ever imagined possible.

Early in her career, Horowitz's hunch was that play had something more to tell us about what dogs might be capable of – that by unpicking this complex behaviour, she might find evidence for theory of mind or something like it. In other words, just as three- or four-year-old children become able to understand that others see the world differently to themselves, dogs might be able to construct an idea of what other dogs are thinking and to exploit this for their own gain: to have as much fun as possible. Horowitz's particular interest was in the moments between moments – the rapid signals and micro-messages that dogs give one another while engaged in exhaustive play. The stuff, in the human speed of life, that it is easy to miss. For Horowitz, these small moments in the everyday interactions of dogs could shed light on the cognitive wheels that turn in the minds of dogs. So she set to work with a video camera, accumulating hours and hours of footage.

'I certainly enjoyed watching dogs playing – the dog I lived with, and all her friends, in particular – but watching play a thirtieth of a second at a time is incredibly laborious, and definitely ate away at that pleasure,' Horowitz tells me. 'Coding videotape – translating what we see into a catalog of quantifiable behaviors – is painstaking work. Moreover, it no longer reads as "play" in slow-motion: play is what we see at normal speed. A frame at a time, one only sees posture, small movements, proximity.'

Key to Horowitz's research were 'play signals' and 'attention-getters'. Play signals include the play-bows, the classic dog 'smile', the high tail that wafts back and forth like a flag on the end of a tall stick. All the classics. Play signals differ from attention-getters. Attention-getting behaviours normally see a dog doing all it can to get into the visual field of another individual, be it human or dog. While engaging in attention-getting behaviours, dogs become, in no uncertain words, impossible to ignore.*

What became clear from the slow-motion video footage that Horowitz captured was that the attention-getting behaviours are vital to keep play going. If one dog loses interest momentarily, the other dog will do all it can to bring the focus back to allow the play to continue. They bark, they whine, they jump into their playmate's field of view, working all the dials to win back their gaze. Then, their attention grabbed, the playful dog deploys the all-important play signal: the play-bow or an enthusiastic bounce with tail-wagging. Horowitz saw a routine to play behaviours: dogs acquire attention first, then send out an invite to play. Interesting? Absolutely. Unique? Not quite. After all, many mammals enjoy multi-faceted and complex modes of communication like these when it comes to play: kangaroo joeys sparring with mother regularly indicate their playful intent using a gentle shaking of the head, which kangaroo mothers give back; voles produce a play 'scent' to show others their desire for fun; seals nose one another around the muzzle and the nape of the neck, nagging friends and family into action. But there would turn out to be an extra layer of complexity in the slow-motion interactions of the dogs that Horowitz studied.

Horowitz found that dogs employed different strategies depending on how inattentive their playmates had become. If a playmate became momentarily distracted by something else, such as a passing bird, a dog might get in the face of its playmate or leap backwards theatrically while maintaining eye contact. But if a playmate became very distracted, such as by another

---

* Dogs use other 'attention-getters' – they nuzzle or push their bodies against the object of their desire or employ a tender paw to nudge you into action. Barking, of course, helps. In fact, many dogs readily employ barking when they see other dogs playing.

approaching dog, then the dog altered and adjusted its behaviour from normal barking to 'HEY! BUDDY! OVER HERE!' levels of intensity. In this way, dogs weren't just playing; they were *managing* play. They were altering their play behaviours with their audience seemingly in mind. Their behaviours weren't deployed randomly but by apparently considering what others might require. In this way, when engaged in play, it is like a dog is directing a pantomime with an actor who doesn't have a memory for lines and has eyes that wander all over the set, taking in the curtains, the props, the instruments, *etc. Connect with your audience*, an attention-getting dog will urge of its lead. If that doesn't work, like a frustrated director considering how best to get the most from her cast, the dog must try a different approach.

Not all dogs are adept at this kind of mind-reading, of course. Just as with humans, some dogs are highly social – they work the cocktail party, so to speak – while other dogs blunder around barking wildly as if they have never seen another dog in their life. This makes uniform statements like 'dogs display theory of mind' a little unwieldy. This is a criticism that Horowitz does not shy away from: 'Their skill at using attention and play signals hints that they may have a rudimentary theory of mind: knowing that there is some mediating element between other dogs and their actions,' she writes in *Inside of a Dog*. 'A rudimentary theory of mind is like having passable social skills. It helps you play better with others to think about their perspective.'

The beauty of Horowitz's study is that – rather than rely on good cops and bad cops, on captive apes, on glass screens or deceptive tricks – the events between dogs played out in the most natural of ways in the most natural of places, without the need for eloquent ruses or problem-tasks or treatments. You can see it for yourself, upon each and every visit to your nearest park.

Such experimental set-ups as these would have been unimaginable four decades ago, when the academic study of dog cognition was considered, at best, unworthy of study and, at worst, reputational suicide, and when the fuzziness of animal minds made them appear unexplorable, closed-off and distant. But, starting with Hare and Csányi and continuing through Horowitz, Rooney and Bradshaw, the fortunes of dogs in research came back into vogue in a very big way for the cognitive science community. Now the research relationship with dogs was changing again. Instead of performing research through 'tasks' or 'problems', dogs had revealed something of their cognitive hardware through simple acts of fun and games – through running, bounding, tugging, barking, twisting, turning, rolling over and wagging. Through pleasure. Thanks to research like Horowitz's, a rare type of ecological relationship was forming between dogs and human scientists – one in which both players were bettered in pursuing the big questions of our respective species. *How best can I play?* asks the dog. *How best can I know?* asks the human. In time, light would be shed on both questions, but it would take other wonderdogs to convince us.

Looking across the TV studio, there is activity all around. Runners and show assistants set up what looks like a giant ball pit (minus the balls) while camera operators spin and synchronise in readiness for the shots that will come once the adverts have finished. Floor managers come and go. Lights are set. Cameras are staged. Around the edge of the empty ball pit a member of the studio staff lays out what look like hundreds of small stuffed toys and other objects, each one different from the one next to it. A baseball. A football. A fluffy duck. A toy lamb. A tiger. A Tigger. And there, finally, a dog.

The bright-eyed piebald border collie moves in and around the hubbub, his eyes on the lookout for opportunities for fun. He spies the toys and grabs one. Apparently without nerves (in fact, could that be … swagger?), he spots a cameraman who looks like he has time to spare. He trots over with a small soft toy in his mouth. The dog drops the toy at the cameraman's feet and looks up, tilting his head in that familiar way. The cameraman gently kicks the toy across the polished studio floor and it skids off, the dog scrambling after it. At this, an assembled audience of a few hundred people laugh, supported by another few million soon to be watching live from their living rooms. Everyone is falling slowly in love with this dog named Rico, and yet the show has barely begun. The TV programme is Germany's *Wetten das?* (*Wanna bet that?* in English), one of Europe's most successful TV programmes ever at that point. The year is 2001 and Rico is about to become more than just your average Saturday-night superstar.

The premise of *Wetten das?* was simple: during the show, a panel of celebrities is shown a procession of people who claim to be capable of extraordinary things and then they are asked to bet on their capacity to achieve it.* So the celebrities might bet on whether someone could construct a V8 engine in nine minutes, or on whether a blindfolded ornithologist could tell a species of bird just by holding its feather, or on whether a farmer could recognise his cows by the sound produced by their chewing apples. For each challenge, a celebrity panel hears the pitch and then makes their bet. Unbeknown to the panel at this point was that Rico and his human companion were a dead cert. Rico, it would turn out, knew the names of some 200 objects. Through play, he was going to show us.

The live TV light pings on. We are back to transmission. Rico's owner moves to the centre of the large empty ball pit and Rico follows attentively. Bright, zippy muzak plays in the background as the lights go up. The task begins and Rico looks up into his owner's eyes obediently. They share a moment

---

* In the UK, the same show was commissioned in the 1980s under the name *You Bet!*

before she utters the first word: '*Schneemann*.' Rico seems to nod and off he goes. With a spring in his step, Rico does a lap of the outskirts of the pool, eyeing up each plaything. His tail wafts lazily left and right as he runs his gaze over each and every object. And then, at the end of the lap, Rico spots it: snowman. He grabs the toy snowman in his jaws and springs his way back to his owner. She bends down low and does a joyful clap of her hand. '*Prima*,' she says. 'Fine.' The crowd bursts into applause at this but Rico barely seems to notice. He trots around from person to person, gleefully trumpeting his snowman to any humans who might be interested. And they are interested. Really interested. Millions of them, in fact.

'*Wo ist das Pokémon?*' her owner says next. Rico sets off on another circuit, carefully inspecting each and every object as his search for a toy Pokémon (Pikachu) begins. No luck at first. He jumps out of the empty ball pit and inspects the objects from outside it, perhaps to get a better perspective. Still no joy. He begins lap three of the ring of toys. It is almost like Rico is attempting to get the crowd on edge now – as if he knows what gets the ratings. Still no Pikachu. At this point, Rico tries something new. He climbs up onto the lip of the ball pit and carefully treads a path over each toy. The crowd gasp at this. Then, suddenly, an excited lunge at a toy hidden from view. There, in his smiling mouth, Rico triumphantly holds up the Pikachu doll. The crowd roar. They want more! Rico duly obliges them. Next, a toy ball. A roar from the crowd again. *More!*

Rico really seems to be enjoying himself now. He looks up eagerly for his final instruction. To finish off, his owner gives a more difficult challenge. '*BVB*,' she says, referring to a toy football in the team colours of Borussia Dortmund (nicknamed 'BVB' in Germany). This one seems to really throw Rico. He looks slightly uncertain. He circles round twice. Stops. Turns back. Starts again. He searches. Each time he approaches a toy, there is the tiniest intake of breath from the crowd. And then, finally, he sees it – the familiar yellow-and-black ball. The presenter lets out the meekest intake of breath at this moment. Rico grabs it and prances around with it in his jaws, as if he has just scored a winning goal in a cup final. The crowd really roars at this. Applause rings out wildly and, the frivolities over, the credits start to come up. The show ends and Rico becomes a Saturday-night star to millions of households – loved and admired from afar. And noticed by a scientist, watching from home, totally amazed at what she saw Rico achieve. Other evolutionary anthropologists will see Rico's performance too. Their jaws will also drop.

On that fateful Saturday night, Julia Fischer was then a post-doctoral fellow at the Max Planck Institute for Evolutionary Anthropology in Leipzig, Germany. Soon afterwards, she told the story about Rico to her colleague Juliane Kaminski. Kaminski's eyes widened – she too had been made aware of Rico and was duly amazed. The two scientists made contact with Rico's owner and enquired whether they might meet Rico to see for themselves up close his peculiar knack for remembering and referencing the names of objects.

Originally, both Fischer and Kaminski were sceptical of Rico's abilities.* In truth, animal behaviourists often are with feats such as this. The history of animal science is, after all, littered with stories of animal ingenuity or derring-do that later turn out to be wildly misinterpreted. The most famous example is Clever Hans ('der Kluge Hans'), a horse that lived in Germany at the turn of the twentieth century and was widely believed capable of the most incredible feats of mental arithmetic. Hans' trainer could give him all sorts of mathematical

* 'I was hesitant to believe that he could really use the spoken words to distinguish the objects,' Kaminski informs me. 'I was sure he was using the owner's body movements and signals that she was giving unconsciously.'

questions and Hans would think them over before tapping out the correct number with a 'click-click-click of his hoof'. Hans was said to be able to add, subtract, multiply, divide, tell the time, work out fractions, understand months of the calendar, differentiate musical tones and read German. For this reason, understandably, Hans was a sensation. But that would change in 1907, when the psychologist Oskar Pfungst worked out that there was a far simpler explanation of events. In truth, Hans could read, but it wasn't German. Hans could read faces. He was picking up on the subtle behavioural responses of his trainer, or others in the room; the tiniest gasps or changes in the frequency of inhalations and exhalations; the imperceptible changes in the eyes or mouth when the correct answer was reached. The trainer, of course, was unaware of his give-aways (in poker terms, his 'tells') but Hans was not. Hans was clever, sure. Just not in the way that anyone had really appreciated up until this point.

Clever Hans has since become an eponymous character in psychological and ethological circles, a kind of Scrooge's ghost who serves to remind scientists of the so-called observer-expectancy effect. He is a superb example of what happens when natural excitement sees us veer away from the rigorous thinking that Morgan's Canon demands of us. No one was keen to make the same mistake with Rico.

Fischer and Kaminski's first task upon meeting Rico was to confirm that Rico wasn't using the same trick as Clever Hans when locating objects – that he wasn't picking up on the gasps or the wide smiles of the audience or his owner as he happened to sniff at or pick up a particular object. To do this, the pair worked from the owner's house. In one room they arranged a set of ten items while Rico and his owner waited in the next room. The experimenter then joined them, instructing the owner to request from Rico that he bring two randomly chosen items, one after another, from the room next door, out of sight. By sending Rico into the next room, away from his owner, there was no way that Rico's behaviours could be inadvertently influenced. Thankfully, Rico had no problems with the task. Then came a second test: the experimenters would add an unfamiliar object to the others in the adjacent room and the owner would call out a new word, never before heard to Rico. With this, Rico would run off with trademark vim, and come back with the unnamed object. In ten trials, Rico correctly brought back the new toy seven times – a feat comparable to well-trained dolphins, apes and seals. Incredibly, a month later, Rico had retained memories of these new words and could retrieve items based on his previous experiences.

In 2004, Fischer told *The Washington Post*: 'It's like he's saying to himself, "I know the others have names, so this new word cannot refer to my familiar toys. It must refer to this new thing." Or it goes the other way around, and he's thinking, "I've never seen this one before, so this must be it." He's actually thinking.' According to the psychological lingo, Rico was 'fast-mapping' – taking in and responding to new words after only a single exposure. (Humans begin to do this in the early stages of childhood.) Fischer and Kaminski wrote up their findings of Rico's abilities and an international community of cognitive scientists and a swathe of media outlets collectively drank it in.

'When making the point that language learning requires more than just the right environment, psychologists often point out that both a baby and a dog are exposed to language, but only the baby learns to talk. This example may have to change,' wrote the psychologist Paul Bloom in a special op-ed for leading journal *Science*. This really was an incredible deal for scientists the world over: 'For psychologists,' wrote Bloom, 'dogs may be the new chimpanzees.'

Recollecting the significance of the research paper today, Kaminski agrees: 'I think it was a wake-up call for the animal cognition community in general, but for the dog cognition community in particular,' she tells me. 'It showed that a dog, a species that was viewed as a bit of a "dumb wolf" by some, could actually perform a cognitive task that was up to that point seen as something that only primates or large-brained birds could do.'

At the time, some considered Rico to be a kind of canine genius, more of an uber-intelligent freak or outlier than an everyday demonstrator of the cognitive feats that dogs can manage. They saw Rico as the dog equivalent of a cosmological authority on dark matter or an elbow-patched expert on antique crockery from the Edwardian era. But then other dogs like Rico started to come out of the woodwork. Smarter, even more impressive, dogs came forth. It would turn out that Rico wasn't unique in his abilities. He was one of many.

The next star was a collie called Betsy, who got her big break through a *National Geographic* magazine call-out for super-smart dogs. Born in 2002, Betsy could enthusiastically retrieve one of 340 objects by name. She could even work from photographs of those objects, retrieving objects she had never had direct experience of. Her scores were near faultless too. In a single trial that involved forty objects, Betsy could manage scores of thirty-eight. She was a pro. Like Rico, she was an enigma. Her picture adorned the front cover of the March 2008 edition of the magazine, all big bright eyes, slightly tilted head, full-bodied connection.*

Then came a black-and-white border collie by the name of Chaser. Chaser became nothing short of a legend among

---

* Asked about the arresting portrait for the *National Geographic* cover, the photographer Vince Musi recalled: 'I set the lights up, I try to sing to [the animals] a little bit, introduce them to some culture, often Frank Sinatra and sometimes Elvis Costello,' he said. 'And I tell them everything I wanted to do.'

dogs. Where Betsy and Rico were managing a repertoire of words in the hundreds, Chaser recognised a staggering 1,022 toys, frisbees and balls by name. Not only this, but her conceptual understanding of words seemed to have moved into a new gear. She was able to understand sentences, littered with grammatical complexity (such as understanding the difference between, for instance, 'Take the Ball to the Frisbee' and 'Take the Frisbee to the Ball') and she was able to seemingly understand rules – that a fluffy toy can be categorised as both 'Franklin' but also 'Toy' – an example of so-called 'many-to-one mapping'.

In 2013, her trainer, the American ethologist John W. Pilley, wrote in *Time Magazine*: 'For me, the most crucial common characteristic of dogs and toddlers is that they both learn best through play. I made games and other playful interactions with Chaser the basis of an ongoing conversation, speaking to her throughout the day in simple words and phrases just as I would to a toddler.'

Another string in Chaser's bow was her impressive ability to grasp 'Do-As-I-Do' requests. In these activities, the trainer holds the dog's attention and says, 'Watch what I do. Now you do it!' and then the dog follows suit. To Pilley, this was proof that Chaser had a theory of mind, that she possessed a conceptual understanding of the request. That, to perform the task, she would need to put herself in the mind of her trainer. 'When toddlers grasp the concept that we want them to imitate us, science says that they are demonstrating an implicit theory of mind and understand, unconsciously, that another person has a unique point of view different from their own,' he wrote before his death in 2018.

Chaser died less than one year later from natural causes. She was an old dog. Another wonderdog that brought humans and non-humans even closer together. Brains divided not by capacity but by degree. That's her legacy. What Rico started,

Chaser chased. Remembering Rico today, Kaminski tells me: 'I remember Rico as a very friendly, active (but also surprisingly calm) dog. He was enormously focused but it was obvious that he simply enjoyed the game. It was not an obsession. He simply enjoyed doing the job – he was just as happy on a regular walk through the forest.'

The feats that Rico, Betsy and Chaser displayed tell us a great deal about the incredible talents that dogs have for memory, for language and for picking up on even the most subtle of gestures. But, as with Horowitz's studies, it is play that allows us to see it. Happiness. Enjoyment. It is fun and games – that universal mammalian phenomenon – that helped us unlock this new age of science. No sedation, no straps, no electric shocks. Just the beautiful art of messing around.

The great experiment that nature began – to tame a wild wolf in the presence of humanity – gave us one of the most fun-loving animals on Earth. In so doing, we were given all the tools we would need to unlock more secrets of their cognition – to get closer to understanding what it really is like to be a dog: to feel what they feel and to know what they know about emotions, contentment and perhaps even love.

CHAPTER 12

To see love coming

> 'LOVE – what is love? A great and aching heart;
> Wrung hands; and silence; and a long despair.
> Life – what is life? Upon a moorland bare
> To see love coming and see love depart.'
>
> – Robert Louis Stevenson, 'Love, What Is Love?'

I would say that, of all questions I am asked about animals, the question of love and dogs and other pets comes up the most. In stage events, at book festivals or even in classrooms, I know it's coming and that the enquiry will air itself in the questions at the end. I am far from alone in having this question haunt me in public life. Other animal-writers say the same. People want to know – they really, really want to know – do our pets love us? They want to get past the idea that it's just a big ruse on behalf of the dog – that there is cupboard love, but nothing more.

Answering this question scientifically is a dangerous game. After all, love is hard to put into words, which is why we tend to celebrate the poets, scholars, musicians and other artists who get so close. Love is hard to quantify, hard to shove on a graph and hard to calibrate. Asking a scientist to define love is like asking a child to describe the colour red. It's, well, it's just … red. Sure, we could drill down into the specific wavelengths of light and we could compare those wavelengths to other colours and discuss how they differ in their peaks and troughs, but we can't express cleanly what is clearly in front of our eyes: red. The reflected light from a rosy apple. The reflected light from a cricket ball. A cherry. A rose.

Yet, as I consider the incredible scientific journey that dogs have guided us through in the last 150 years, I find myself increasingly certain about this powerful attachment. Through the discoveries made in the last ten years, particularly, I feel we are now able to make a scientific case for this kind of attachment in a way that subscribers to mainstream science could acknowledge and accept as something close to truth. Line by line in this final chapter, using Robert Stevenson's titular poem, I lay out my case for love, with the evidence acquired from the current crop of scientists, each inspired by recent generations of remarkable wonderdogs.

A great and aching heart

One of the themes of this book is how, like a virus, scientific ideas move from country to country, propelled by journals, by books, by press, by people. Scientists play no small part in this spread. They infect seats of academia as they travel. Their perspectives seed themselves in campuses and their corridors, in the minds of colleagues, friends, acquaintances.

Pavlov's ideas – especially his theory of classical conditioning – was one of those viruses. It was a virus that would eventually find its way to the United States via a carrier – an academic by the name of W. Horsley Gannt (1892–1980) who, in 1929, moved from Pavlov's laboratory in Russia to Johns Hopkins University in the USA. There, Gantt founded the Pavlovian Laboratory in his honour.

Gantt saw in dogs a perfect research animal with a 'special advantage 1) stemming from his long and intimate association with the human being, and 2) because of his very responsive and easily influenced cardio-respiratory system.' Gantt was known for many discoveries* and published more than 700 articles in his lifetime. As well as developing our knowledge of how Pavlovian conditioning responses play out in the everyday physiological responses of animals, he also made great strides in other early psychological fields, including human neuroses and exploring the impact that mind-altering drugs have on mood and sensation. Many of his experiments looked at 'cardiovascular conditioning' – how heart rate and blood pressure can be influenced by external conditions or factors. A classic example of this phenomenon occurs when athletes train on mountainsides and, due to lower atmospheric oxygen levels at high altitude, see their number of red blood cells increase over time.

This focus on heart rates gave Gantt a special insight when it came to dogs. Gantt and his team noticed something of a cardiovascular oddity when dogs and humans met one another. He found, as one example, that when a human walked into the room, a dog's heart rate would spike before gently tailing off to a more normal rate. Gantt also noticed that if a human was given an opportunity to stroke this given dog, the dog's heart rate would lower more quickly than if the dog was left alone, untouched. Gantt's studies suggested that a dog's heart would best

* The Gantt chart was not one of them. That honour went to Henry Laurence Gantt (1861–1919).

slow when petted by a special subset of people: people with whom, in the words of Gantt, the dog had a 'special relationship' – they were companions, friends, family. The obvious and repeatable response in heart rate that he observed was given a name: dogs were showing, in Pavlovian terms, something he called the 'social reflex'. Their body processes could be affected by the presence and physical interactions of another species – us.

Gantt's studies were the first to demonstrate that the emotional states of dogs and humans could be mapped. They could be plotted on a graph, compared and contrasted – studied empirically, in other words. For the first time, the physiological responses to dog and human companionship could be put to paper. The social reflex could have been a fertile ground for new insights into the human–dog relationship yet, sadly, the idea never really took off.

As psychologists looked to dogs for insights into depression and how to fix it, and as geneticists looked to dogs to understand the intricacies of their development, Gantt's ideas on this issue went into a kind of stasis. Later, after the cognitive revolution, they would come back in a different (though familiar) form. Instead of measuring heartbeats, this new breed of physiologists would look to chemicals instead. The *y*-axis they favoured was a molecule produced by the brain called oxytocin, a hormone that spikes when humans partake in pleasurable activities. Oxytocin is especially important to female mammals – it surges, particularly, during the nursing stage – but male mammals also produce it during moments of tenderness. Bluntly, in humans at least, when levels of oxytocin rise, we see behaviours of love; when it plummets, we see war. Oxytocin (time-honoured tradition hereby sees me refer to this molecule at least once as 'the love hormone'*) has become a go-to

---

* Many scientists absolutely hate this term. As Shelley Taylor of the University of California, Los Angeles, puts it: 'It's never a good idea to map a psychological profile onto a hormone; they don't have psychological profiles.'

hormone for those interested in emotional states. It is easy to test for and its traces in blood or urine can be mapped on a graph in a similar way to Gantt's early studies on heart rates.

I first came upon this incredible molecule courtesy of research into the sex lives of animals for my previous book, *Sex on Earth*. I got to learn first-hand from scientists that, by manipulating the concentrations of oxytocin in the brains of male prairie voles, for instance, one could change them from monogamous family guys into something like licentious lounge-lizards. I had learned that all mammals depend on brain chemicals like these to reward behaviours such as suckling, grooming, play – activities that serve, unknowingly, to propagate genes into future generations. Hormones that literally give the brain a little buzz: a 'I-should-do-that-again-sometime' kind of kick. Could this be the means through which we explore the love that dogs and humans feel for one another? Possibly, argue some scientists.

Takefumi Kikusui, a behaviour scientist at Azabu University in Sagamihara, Japan, understood the role that oxytocin played in bonding human mothers to their babies and vice versa. Likewise, Kikusui knew that oxytocin production could be induced simply by mother and baby staring into one another's eyes. Naturally, he wondered if the same was true between humans and their dogs. In 2015, Kikusui and colleagues set to work to test out their hypothesis.

On the whole, the methodology they devised required very little of its canine participants. Along with his colleagues, Kikusui recruited thirty dog-owning friends and neighbours and invited them to the lab. First of all, both dog and owner were asked to urinate in a cup – a simple means of acquiring data on oxytocin concentrations. Then they were sat down in a room together for thirty minutes, during which time they were allowed to cuddle, stroke and (importantly, it would turn out) gaze longingly into one another's eyes. Their thirty minutes up, it was time for another urine test and, with that, the participants were on their way.

The results of this study were profoundly telling: those owners and their dogs that gazed into one another's eyes were dosed up to the nines on oxytocin in a way that those who barely met one another's eyes were not. Of those individuals who maintained the most eye contact, the dogs saw a 130 per cent rise in oxytocin levels during their half-hour alone with their owner. The human owners were even more profoundly affected: some saw a 300 per cent increase in oxytocin in the same period. Eye contact really seems to matter in the human–dog relationship.

In the same study, Kikusui and colleagues also sought input from individuals who were raising wolves, rather than dogs, as companion animals. In the experiment, wolves rarely met the eyes of their keepers. (This finding came as little surprise: to wolves, eye contact is a threat rather than a sign of connection.) Neither did the wolves show any surge in oxytocin levels. Ours is a different relationship to the one we share with dogs, they deduced.*

Unsurprisingly, the study drew interest from a number of quarters. Brian Hare (he of 'my dog does that!' fame) declared in *Science* that it was 'an incredible finding that suggests that dogs have hijacked the human bonding system'. In *National Geographic*, Larry Young from Emory University (part of the prairie vole study mentioned earlier) said: 'During dog evolution, we have probably selected for a behaviour in dogs

* In a subsequent test, Kikusui looked at the effect of spraying oxytocin solution via a quirk squirt up the nose of both dogs and owners. Interestingly, this saw gazing interactions increase by 150 per cent between female dogs and their owners. Male dogs, however, remained unchanged for the duration of their loving gazes. Kikusui suspects this is because females are more sensitive to oxytocin, given that it plays an enormous role in lactation, labour and reproduction.

that elicits a physiological response in us that promotes bonding … That behaviour is eye-gazing.'*

When we look into each other's faces, we are flushed with the same chemicals: warm feelings, connection, attachment. We have no reason to suspect that these chemicals work differently on them, or any other mammal for that matter, than on us.

Is it love, then? On its own, not quite. For our notion of shared love to be more resilient to volleys from the critics, we're going to need much more than that. In our search for love, we have further pillars of truth to erect …

Wrung hands; and silence; and a long despair

While researching this book, I have noticed two camps among the dog writers. I notice the freedom with which some use the L-word, and the winces this word evokes in others – the more militant disciplinarians of science (including those who used to be my teachers) who find the use of this human word completely unsuitable when applied to animals other than humans.†

For this reason, rather than open themselves up to criticism from peers, many biologists prefer the term 'attachment' than love – a word often used to describe relationships that animals manage in order to maximise the flow of their genes into

* When engaged in a session of mutual eye-gazing, dogs can also 'catch' a yawn off their owners. In doing so, they join an esteemed group of animals that exhibit contagious yawns – just humans, chimpanzees and baboons manage the feat.

† Even the strictest zoologist cannot escape the fact that parts of our language, used by scientists and laypeople alike, is imbibed with meaning. The word 'puppy', for instance, comes from the French *poupée*, which means 'doll' – an object cherished and adored by children.

future generations, be it through sex, through survival of family (kin selection), of offspring or of the offspring's offspring.

It is clear that dogs feel attachment to us. Darwin understood it; so did Romanes and Lubbock. Scott and Fuller showed it clearly, by preventing and restricting periods of human–dog connection during puppyhood and noting the damaging result. In 1998, Csányi and Miklósi (of the Family Dog Project) proved it beyond doubt, putting dogs through a psychological test normally applied to human infants and noting that the resulting distress was directly comparable to that seen in child–parent attachments.

John Bradshaw, author of *In Defence of Dogs*, has been consistently vocal on the issue. As the dog population continues to rise, his concern has been for a long time that cases of separation distress will rise with it. Depending on the breed, manifestations of separation distress include biting and pulling apart furniture, barking, whining, howling, urinating, defecating and even vomiting. Self-mutilation can also occur. Bradshaw's research suggests that separation distress is far more serious an issue than we once thought. In his own research, including one longitudinal study in which the development of forty puppies (from twelve dogs in total) were logged over a period of eighteen months, Bradshaw found that approximately half of all puppies went through a stage of significant emotional distress when left alone, peaking at about twelve months. Subsequent questionnaire studies set more alarms ringing for the research team. In 676 interviews, 17 per cent of dog owners reported that their dogs were showing regular signs of distress at being left alone. A further 18 per cent reported periods in the past where such distress occurred and had been resolved. The implication of Bradshaw's research is, quite frankly, concerning. The UK has around 10 million dogs and the number is rising – if one-fifth of these dogs are regularly suffering separation distress it means that, on a given day, two million dogs might be suffering anxiety, stress and trauma at our not being there. Scaling that up to include all of the planet's dogs (removing the vast majority of dogs that are unrestrained), a conservative estimate of dogs that suffer separation distress each day puts the number in the tens of millions.*

Bradshaw's opinion is that separation distress is not a disorder but in fact is what should be expected from any relationship between two highly social, bonded, cognitively endowed animal companions – including between adult humans and children when they are separated. 'We have selected dogs to be highly dependent on us, so that they can easily be made obedient and useful,' he writes. 'Why is it so surprising that they do not like being left alone?'

The strings that bring about these emotional responses are brain hormones, including oxytocin. They are demonstrably the same hormones that emotionally attach puppies of both wolves and dogs to their mothers, or ourselves to our children, and us to our parents. In this way, through the branching boughs of shared ancestry, the cogs of our attachment have come from the same deep-rooted place as theirs. From mammals that lived long ago and whose lives played out in the same language of DNA, written in the same unshakable amino acid font: adenine, cytosine, guanine, thymine. Not L.O.V.E. but four letters still: A, C, G, T.

Life – what is life?

We are halfway through the final chapter and so it feels like a fitting time to consider the nature of the perennial discussion about whether, and to what degree, dogs love us. Perhaps you

* For those seeking advice on appropriate company for dogs, the RSPCA's advice is, as ever, excellent: www.rspca.org.uk/adviceandwelfare/pets/dogs/company.

are sure they do. Or perhaps you remain a sceptic on the idea. If you are the latter, what would it take to convince you? Here is a metaphor that summarises my own perspective on what it took to convince me.

As I write these words, scientists are arguing over whether there is life on Venus while members of the public (including me) are watching from the sidelines, rapt. The implications of this research are truly staggering. After all, if life is found on Earth and next door on Venus, the chances are very high that it will exist elsewhere in the galaxy and, almost with 100 per cent certainty, elsewhere in the universe. For me, if true, a childhood dream (buoyed by Spielbergian sentiment) would be realised.

But why Venus? Why are we talking about this now? The recent drama about life on Venus has come courtesy of the discovery of phosphine gas in the planet's upper atmosphere. On Earth, phosphine is produced by microbes. The contention is that its presence in the Venusian atmosphere suggests there are microbes on Venus as well. It's an exciting time to be alive but, alas, the shaping forces of scientific enquiry are (as I write these words, at least) going to work. Right now, the opposing scientists argue that the original research team's readings are wrong. They highlight that the assumption that phosphine is a molecule that doesn't naturally occur without life could be false. And, in the main, these critics should be celebrated for attempting to find the cracks in the argument. Science is, after all, always better for debate.

So what would it take to convince these sceptics that life exists on Venus? More readings of Venus's upper atmosphere would help, certainly. But other sources of evidence would also be required to convince scientists that life really is present there. Methane, perhaps. Oil, maybe. Fossils of nodules that suggest early bacteria might help change the minds of some critics. But some would need even more evidence if they were to be silenced on the matter. They might need DNA. A photo. A specimen. If all these things and more could be acquired, one would have a hard time convincing the sceptic that there was no life on Venus. It would be like walking into a room of clocks, all about to strike twelve, and denying the concept of midday or midnight. Only a fool would dare.

I believe, having come to the final chapter of this book charting 150 years of science, it is the same with humans and dogs. Our love is apparent not from the strength of a single research paper but from the strength of so many corroborating sources that have been hit upon in recent years. So far we have named three: heartbeats, brain hormones, separation distress. But there are more. Plenty more.

Upon a moorland bare

In 1914, quarry-workers in Oberkassel, a suburb of Bonn in Germany, discovered something remarkable. A host of archaeological discoveries in the previous century had shed light on early human burial practices including the burying of humans alongside prey, but the Oberkassel site was different: among the human remains the quarry workers discovered was the body of a person buried alongside the remains of a small dog. For such a prehistoric site, the discovery was somewhat startling. The 14,000-year-old remains would become the earliest evidence we have of a human–dog burial – an unmistakable symbol, written in bones, of a loving relationship between one species and another.

The jawbone of this prehistoric canine (the so-called Oberkassel Dog) continues to excite scientists in the modern day. One scientist particularly tickled by the fragment in recent times is Leiden University's Luc Janssens, who has something of a symbiosis all of his own. Straddling both the fields of veterinary science and archaeology, Janssens and colleagues saw in this single remaining 14,000-year-old jawbone an opportunity to assess the health of the dog that once owned it.

Upon closer inspection, the research team spotted something previous generations of archaeologist had not noticed: the teeth were worn and had not developed properly. The dog had suffered disease, they deduced. He or she had been ill. In life, judging by the condition the mysterious dog was in, they would have had to have been cared for by humans to have survived for so long. The disease that the team discovered evidence for was not rabies but another virus that frequently washes through dog populations: canine distemper. In puppies, one noteworthy symptom of canine distemper is damage to the enamel that forms on developing teeth. According to Janssens' analysis of the jaw, the 28-week-old puppy most likely contracted the disease ten weeks beforehand and may have suffered bouts of serious illness that would have lasted weeks at a time.

'Since distemper is a life-threatening sickness with very high mortality rates, the dog must have been perniciously ill between the ages of 19 and 23 weeks,' Liane Giemsch of Archäologisches Museum Frankfurt, the paper's co-author, told *National Geographic*. 'It probably could only have survived thanks to intensive and long-lasting human care and nursing.'

The Oberkassel Dog is one of a number of archaeological dog specimens that are being revisited by a new wave of scientists with innovative tools at their disposal, itching to revisit the mystery of how humans and dogs first met upon the moors of time. Armed with knowledge about disease, veterinarians like Janssen and Giemsch are helping. But alongside them, playing an increasingly important role, are geneticists, swab in hands, eager to sample the genomes of dogs and learn new secrets about their past. Like Janssen and Giemsch, these scientists can tell new things from old bones. And then some.

Among the dog genes that have evoked much interest in recent years are two that go by the name of GTF2I and GTF2IRD1. In both dogs and humans, 'insertions' or 'deletions' upon these genes (and others nearby) seem to bring out an effect on personality, influencing how sociable individuals might turn out to be. In humans, accidental deletion of these genes causes Williams-Beuren syndrome (WBS), a genetic condition that causes symptoms including delayed development and physical and mental health problems, as well as a charismatic tendency to be talkative and excessively friendly. It increases the likelihood of a so-called 'cocktail party' type of personality.

In dogs, insertions upon these genes bring on something similar, most notably 'hyper-sociability', a condition in which dogs show even greater regard for people, other dogs and, well, everyone and everything. Hyper-social dogs are often full of joy, topped up on life, eager to please and eager to connect. But they can also suffer, to a greater degree, the downsides that love brings: separation distress, anxiety, howling, wailing.

In humans, mutations across this key collection of genes are very rare. About one in 18,000 people are born with Williams-Beuren syndrome. In dogs, however, their genes are riddled with insertions across these crucial regions. This finding suggests, quite clearly, that natural selection has favoured those prehistoric dogs that were most social – willing to engage with, to work with, to connect and engage with humans. These hyper-social dogs are more common than you might think.

'The average dog carries two to four of these insertion mutations, with some breeds (or groups of breeds) carrying far fewer while others can have many more,' Bridgett vonHoldt, an evolutionary geneticist at Princeton University, informs me. 'It is rarer, but not impossible, to find dogs that carry more than six mutations.'

As luck would have it, vonHoldt's own dog – a gorgeous sheepdog by the name of Marla – scores a four. VonHoldt knows this because (who hasn't?) she has swabbed and analysed Marla's genes herself. Marla's high score explains what comes across in many of the photos of her: she looks like a dog whose every waking moment is taken with togetherness. Indeed, according to vonHoldt, she has become a firm favourite on campus with passing students requiring a pick-me-up to get through exams or other hard times, regularly coming over for a stroke. 'I know that Marla is only content when her humans are together, and she loves interactions with humans with every cell in her body,' vonHoldt wrote. 'She is such a fascinating creature, with such a desire to be with her humans. I always ponder the major changes that had to have happened to domesticate a wolf into a dog like Marla.'

Wolves have, on the whole, about half the insertions across these genes as dogs. This suggests that selection pressure on sociability must have occurred after the dogs' split from the ancestors of today's grey wolf. If true, their gregariousness waves like a genetic flag, hoisted by the interactions of our two species many thousands of years ago at a moment when dogs began sprinting off down a different evolutionary path to wolves – one where the friendliest, rather than the fiercest, prospered.

According to Clive Wynne, founding director of the Canine Science Collaboratory at Arizona State University in Tempe, USA, this hardwiring for sociality explains many interesting facets of dog behaviour. How, when captive wolves and dogs are put in a space with their caregiver, dogs appear to spend four times as much time up close and personal with that caregiver. How, if you allow a dog to run through a door into a Y-shaped room – a room in which food is placed in one arm of the 'Y' while in the other is the owner – most dogs will invariably go towards the owner rather than the food. That when human caregivers feign being stuck in a box while a dog observes, in most cases the dog clearly shows signs of distress, crying and whining and pawing at the box in an apparent bid to help the caregiver escape.

Dogs want to be near us, Wynne argues in his book *Dog is Love*. In fact, if you have a dog, the evidence is probably playing out right now, as the dog lies at your feet or watches you from the sofa. Though this seems obvious, a great swathe of studies prove what is (literally, sometimes) staring us in the face. In his public appearances, Wynne calls himself 'a reluctant convert' to the idea that dogs love us. Originally, he considered the close bond that dogs and humans have as something more clearly explained as attachment, a relationship influenced by our being their primary means of nutrition, but over the years the studies he has undertaken, along with those of his students and colleagues, have forced him to reassess this opinion. He needed a nudge, of course. The arrival of his own dog Xephos, a lovable whirlwind of mutt-kind by all accounts, appears to have been a catalyst in his final conversion.

'We love her. She loves us. Actually, she loves almost everybody,' Wynne explained to an assorted online audience at the 2020 International Society for Anthrozoology Annual Conference. 'She very, very rapidly makes these strong, powerful connections with people,' he smiled at the online audience (myself included) cooing in response.*

* For those not in the know, 'anthrozoology' (sometimes called human–nonhuman-animal studies, or HAS) focuses on interaction between humans and other animals – it's a kind of wheel-house where anthropologists, physiologists, psychologists, veterinary professionals and zoologists meet to discuss and exchange thoughts on research, with the ultimate aim of improving human–animal relationships for all parties, us and them. I heard Wynne speaking at the twenty-ninth meeting of the Institute for Anthrozoology, organised by the University of Liverpool. I had a great time. Register for future meetings at www.isaz.net.

I was rather surprised to see how freely Wynne used the L-word at this and other speaking engagements. He has, after all, a reputation for science, for rational thought and considered debate. He is not someone who has ever shied away from being critical of anthropomorphic rhetoric. In the past, Wynne has been a vocal critic, for instance, of Brian Hare's pointing studies that suggest that dogs have innate cognitive skills to understand gestures that have evolved specifically with humans in mind. Wynne argues, instead, that dogs can apply similar logic to other animals, including penguins and pigs, if their life experiences – their socialisation period – plays out differently than it does in most human homes. Wynne's unrelenting questioning of some of the attributes of dog brains has seen him labelled in the most unflattering ways by some: one author referred to him as 'psychologist Clive Wynne, a former pigeon researcher considered by some to be the "Debbie Downer" of the canine cognition field'. And yet, there he was in these speaking engagements, like some sort of authentic, rational, considered yet doe-eyed, lab-coated hippy – all sweetness, superpowers, love. Xephos, the power of a wonderdog, changed him.

Hearing Wynne talk so openly about the powerful attachment that he and Xephos felt for one another reminds me of the attachment I feel, along with my family, right now for our own dog – a lurcher by the name of Oz. It's a classic dog-owner vibe that Oz and the rest of us share, but that makes it no less remarkable. Quite simply, at any time of day, Oz appears to have a laser-like knowledge about where, in the vicinity, I am. A simple analysis of where his eyes are pointing or in which direction his curled-up body lies while sleeping tells you how much of his cognitive bandwidth I use up. If he could speak, he could tell you when I last went to the toilet, at what points I went upstairs, at what points I came back, when I disappeared to take a call for a moment.

He could tell you how to read my quick, barely noticeable body movements that suggest play is imminent – how I pull the washing out of the washing machine, put it on radiators, empty the dishwasher – the small events are the precursor to our play each morning. Diary-wise, Oz knows everything about me – more than anyone else in the world. This connection really does run deep. Oz reads things in my unconscious movements around the house all day that, if anything, besmirches humanity's reputation for all-conquering consciousness. What does it say about our apparent status as rulers of the world that an animal that lives at our feet can see in us patterns of routine that we never even thought existed?* The truth is, I don't know. I can be clear, though, that Oz has been nothing short of an inspiration for me – he's amazing. He really has been a life-changing part of our existence, much like Xephos has been to Wynne. You will know it too, I'm sure.

Our childhood dog, Biff – a somewhat neurotic Shetland sheepdog with a capacity to be both confident and secure in his own special brand of haplessness – was the same. But Biff just … lived for us – the sound of my feet walking down the stairs each morning brought out in him an excitement close to feverishness. It was borderline convulsive. On my return from school each day, he would joyfully lie on the floor by the sofa, wagging his tail, knowing I would first look for the TV remote and then lie with him, checking football transfer news on Teletext. In those days, walks with Biff weren't about exercise; it was like he saw them as a joyful way for us to channel a mad kind of delight. It was like they were an

* You'll notice that this relationship, on the whole, differs from the one humans share with cats. If one thing separates the two species, it's the hour-by-hour contact – the moments between moments that we share. Cats just aren't interested in the minutiae of our existence. On the whole, many dogs are.

expression of our companionship, laid bare for other dogs and humans to see.

'Love is dogs' secret superpower,' is how Wynne puts it, describing the dog–human relationship in *Science Focus* magazine. 'It was Xephos who taught me what almost every dog owner already knows: it is their unbounded capacity for love, not any distinctive form of intelligence, that singles dogs out.'

Dogs and humans are hardwired to connect; many of us do so effortlessly. In many ways, Marla and Xephos prove it. Yet again, our greatest leaps towards understanding the cognitive and emotional feats of dogs come from those scientists touched, enamoured and drawn in by individual dogs with whom their lives have entwined. They are two of many dogs in this book to show us the way. They join the established ranks of wonderdogs – dogs like Oreo, Flip, Rico, Chaser, Betsy – whose insights have helped us further understand the dog–human condition. Each informed and informs its human companions of something new about the animal world. Each instilled within humans something like awe in everyday behaviours. Each was an engine of science, oiled through play. Significantly, each was loved, cherished and adored. And each gave that intense bond back, in spades.

To see love coming

A hundred years after Gantt spotted the strange effect that a human caregiver entering a room had on a dog's heart rate, another scientist considered how best to study a different physiological response to our two species meeting. The scientist in question considered how exactly dog brains change when they see us. To find this out would take planning; it would require more than just a simple blood pressure monitor and a notebook. To understand what was happening in the brains of dogs would require a dog to sit patiently in an fMRI scanner, an unfamiliar tight space at the far end of noisy human technology. You'd be forgiven for thinking that no dog could patiently endure such a place. But not all dogs are like Callie. And not all scientists have the patience of Emory University's Gregory Berns, the esteemed American neuroscientist and psychologist.

Callie, a black-and-white mixed breed, would become a pioneer for Berns' research – the first of many dogs to take a trip, willingly, into a noisy fMRI scanner to have a brain scan. The first to be trained to do so, apparently happily and with no stress. Callie was Berns' muse – his star pupil.

'Callie's excitement was infectious,' Berns writes in his engaging book, *How Dogs Love Us*. 'Everyone in the lab wanted to see the experiment we were about to perform, mostly because nobody thought it would work. Could we really scan a dog's brain to figure out what it was thinking? Would we find proof that dogs love us?'

To undertake the fMRI sessions, Callie needed training. She needed to get used to the machinery. Crucially, she needed time. To help Callie get used to the idea, Berns (working with a dog trainer) built a mock-fMRI simulator at home. He recorded the sound that the magnets in the scanner made when the machine was running and he played these sounds during games and general play-time with Callie in and around the mocked-up device. In the early days, the sounds were low. Then, a week in, Berns turned up the volume a little bit. Then more. Then more again. Within three months, Callie was accustomed to the rumbling noises generated by the machine. She was ready to enter a real scanner. It was a task she duly completed, her tail wagging.

Callie was a proof of concept for a larger scientific study that would require not just one but a number of dogs to be trained to enter the fMRI scanner. To find canine volunteers for this stage of their research, Berns and his fellow researchers put out a call to local dog-training schools, to friends and family, and to dog-lovers in the local area. Against all odds, they got their responses. While many Americans headed to church on a Sunday morning or cleaned their cars or mowed their lawns, a clutch of twenty or so dogs and their owners would meet up for fMRI training near Berns' laboratory. Just as with Callie, these dogs would take home with them their very own fMRI mock-ups, complete with loudspeakers.

Images in research reports show these research dogs sitting or lying patiently in the scanner room. They have names like Kaylin, Pearl, Eddie, Ohana, Ninja and Zula. Perhaps tellingly, one of the dogs is named Zen. There are other photos of the dogs online. One image sees the dogs gathered around an fMRI machine, staring proudly down the lens. The front row of dogs sit in a line, like a cup-winning football team about to pick up their trophy and shake it around in jubilation. Behind them, another dog stands regally, like a captain, a picture of self-importance, upon the bed that slides in and out of the scanner.

Once trained to happily use the scanner, top of Berns' list of things to explore in dogs was the brain's reward system – the pathway that regulates motivation, reinforced learning and other cognitive processes associated with feelings of pleasure. Of particular interest to Berns was how developed the dog's caudate nucleus would be. In humans, the caudate nucleus is packed with receptors for the neurotransmitter dopamine – a molecule long associated with feelings of pleasure and contentment and, particularly, satisfaction. No one was quite sure whether dogs might have the same structure or how developed it might be. Under his tutorage, the volunteer dogs entered the fMRI machines and, at a certain point, were given a signal to indicate they would be receiving a food reward once the experiment was over. The results were pored over by Berns and his research team, who were amazed at what they saw.

In those fleeting moments, when the idea of a reward was introduced, dogs' brains underwent a frenzy of activity in the caudate nucleus, mirroring that seen in human brains under the same conditions. It was proof that the caudate nucleus was there. That it was lighting up. That it was capable of firing up like our own.

Having discovered this, Berns and his team of experimenters tried a different test. Instead of being offered edible treats, this time the dogs used in the experiments were offered another kind of reward: they were given the sudden thrill of an unexpected appearance by their owner exclaiming excited sentiments. In this test, most (but not all) dogs were apparently as delighted at seeing their joyful owners as they were about the food. Interestingly but perhaps not surprisingly, fragrances also worked. One odour consistently evoked a response in a dog's caudate nucleus – it was the familiar smell of the dog's owner.

The relationship between dogs and their human companion was explored in other ways using fMRI brain scans. For instance, the brain patterns of dogs whose owners temporarily walked out of the room showed a comparable surge in activity when their owners re-entered the room, in a similar way to Gantt's studies into heart rates a century earlier.

Other research institutions have since taken to training their own batches of volunteer dogs to sit in fMRI scanners patiently, in much the way that Berns and his colleagues managed. Subsequent studies courtesy of (who else?) the Family Dog Project, for instance, that involved playing 200 different sounds (including car horns and whistles) to dogs, showed that the brains of dogs lit up in the same way that human brains did when hearing human noises, including human crying and laughter. This discovery, in particular, was met with great interest from cognitive researchers. 'Finding something like this in a primate brain isn't too surprising – but it is quite something to demonstrate it in dogs,' Professor

Sophie Scott (of the Institute of Cognitive Neuroscience at University College London) told the BBC at the time.

The fMRI data, explored now by scientists in the USA and in Hungary, suggests what Darwin, Romanes, Griffin and many modern-day neurologists, physiologists and ethologists have always argued to be true: that human and dog brains differ only by degree. That, in Darwin's words, 'There is no fundamental difference between man and the higher mammals in their mental faculties' – that apparent differences between humans and non-humans are 'of degree, not of kind,' and that, in Lind-af-Hageby's words 'a responsibility rests upon us to see that these creatures, who have nerves as we have, who are made of the same flesh and blood as we are, who have minds differing from ours not in kind but in degree, should be protected.'

This recent fMRI research into the minds of dogs, forms another pillar in our argument that the strong attachment we feel for one another is like the attachment dogs can feel for us. The fMRI evidence slots in alongside oxytocin and heartbeats. Alongside attachment theory. Alongside our discoveries about how dogs are drawn to people, how they gaze our way, how they understand our movements. Alongside the archaeological finds. Alongside our knowledge of genes and their impact of sociality. Alongside the way dogs play, the way they respond and the way they act around us.

That dogs feel love like we do is no longer an outlier's opinion. That their hearts don't beat like ours, that their brains don't light up like ours, or that the warmth you give them is unlike the warmth they give back – opinions like these are now, after 150 years of scientific reasoning, drifting out of the mainstream.

Like Clive Wynne, I have been accused of being a bit of a 'Debbie Downer' on the issue of animal emotion. But I must confess, in the process of researching and writing this book, I have come to see that my logic was faulty. That my assumption on this issue was debasing. Demeaning. Morgan would no doubt roll his eyes at my logic, but perhaps Romanes (and later Griffin) would not. I am certain that some quarters of the science research community will roll their eyes at these final pages, accusing me of a wave of sentimentality now that our journey through the decades is nearly over. But this feels like the right time to say it. To these doubters, to these hecklers, to these nay-sayers and hardline sceptics who question if dogs can love humans like we love them, I guess I can only say to read each line of Robert Louis Stevenson's poem again, slower this time. With each sentence, think of the scientific discoveries made in the last ten years, courtesy of hundreds of incredible dogs working closely with their companions, in a climate of love and compassion, never torture.

LOVE – what is love? A great and aching heart;
Wrung hands; and silence; and a long despair.
Life – what is life? Upon a moorland bare
To see love coming and see love depart.

If you deny dogs the emotion of love, then you must deny the same emotion in one of Scotland's finest writers. Gracefully, warmly, joyfully – I give in.Epilogue … and see love depart

We had all the hallmarks of love for one another, Biff and I. Each morning, when I came down to breakfast, he would jump up with something verging on rabid, delighted hysteria. Every day he would lick my hands and face and shoulders with glee, often toppling me down to the ground. We would lie together, each of us nudging, cajoling, rolling, playing. He would play-bow and I would respond. Then, alongside us as we watched TV, he would sleep contentedly. It was love, no questions asked.

But there is tragedy to all love, and my relationship with Biff was no different. The great tragedy of dogs and people is that our lives move forward at different speeds. Biff's entire life history fitted into only a chapter of mine. At this point, I would really like to tell you that he had become an old dog with a wise and spiritual-leader kind of character. With a grey beard, a knowing wag of the tail, warm eyes, a familiar smile. But he wasn't. In his senior years, Biff developed something of a repressed fury about his status in the world – an angry, begrudging obsolescence to all other life forms, especially hedgehogs. He became surly and tired. Abscesses had seen his teeth removed. His barks, when they came out, were slow and laboured, with long gaps between them.

Biff remained dozy for most of the time in that final year and his brain was clearly fading. Sometimes, he would completely fail to recognise us, not even the smell. Other days, he would want to go on short walks and then decide, midway through, to roll over and give up. Often, these situations would see us carry him home, our backs put out by his sudden incapacity to judge distance.

Epilogue ... and see love depart

We had all the hallmarks of love for one another, Biff and I. Each morning, when I came down to breakfast, he would jump up with something verging on rabid, delighted hysteria. Every day he would lick my hands and face and shoulders with glee, often toppling me down to the ground. We would lie together, each of us nudging, cajoling, rolling, playing. He would play-bow and I would respond. Then, alongside us as we watched TV, he would sleep contentedly. It was love, no questions asked.

But there is tragedy to all love, and my relationship with Biff was no different. The great tragedy of dogs and people is that our lives move forward at different speeds. Biff's entire life history fitted into only a chapter of mine. At this point, I would really like to tell you that he had become an old dog with a wise and spiritual-leader kind of character. With a grey beard, a knowing wag of the tail, warm eyes, a familiar smile. But he wasn't. In his senior years, Biff developed something of a repressed fury about his status in the world – an angry, begrudging obsolescence to all other life forms, especially hedgehogs. He became surly and tired. Abscesses had seen his teeth removed. His barks, when they came out, were slow and laboured, with long gaps between them.

Biff remained dozy for most of the time in that final year and his brain was clearly fading. Sometimes, he would completely fail to recognise us, not even the smell. Other days, he would want to go on short walks and then decide, midway through, to roll over and give up. Often, these situations would see us carry him home, our backs put out by his sudden incapacity to judge distance.

Biff would sleep for hours at a time on those days – post-breakfast, pre-lunch, post-lunch, early afternoon, late afternoon and, while we watched TV, all evening, curled up at our feet. These moments, I enjoyed a great deal. His paws would clock up imaginary miles. His tongue and jaws would twitch and click. I would watch his eyeballs jerk one way and the other, tittering in their orbits, eyeing up unseen multitudes. Biff was an intense dreamer in his later years, and naturally I would wonder what his dreams were like, where they took him and how they made him feel.* I often wondered whether he was a young dog in those dreams or, in those later years, whether he suffered as much as in his waking state. Now, it is all I can do to stop imagining whether my own final suffering, when it comes, will be the same.

I was away travelling when it happened. I knew instantly when I returned. I pulled up outside the house and saw my mother and father poke their heads out of the front door, and noticed the subtle difference in how wide the door was being held open. Normally, they would be worried about the dog getting out and onto the road but they appeared to have no such concerns now. I knew then that Biff was gone, that his decline had continued and there was no coming back.

There wasn't tears at the news, just the heavy weight of loss. I was sad for him and his suffering in the final year. But I'm a little ashamed to admit that I was also sad for me too. That I wasn't there at the end. That others had been part of his ceremony. That this friend had gone through life at a hearty pace, upon tracks towards death that I too was riding, just more slowly. The grief lasted months.

---

* The area of the brain that paralyses muscles and stops us acting out our dreams is called the pons. It is slow to mature and quick to degenerate. This is why young dogs and old dogs play out their dreams in such a vivid and physical manner.

My biggest regret about that time is that I didn't talk about it more. That I chose not to mention it to my work colleagues or my friends, for fear that they would think me shallow or stupid. I sensed a stigma about what the 'right' amount of grief for the loss of a pet should be. It seemed to me that dogs seemed to fit somewhere on the scale between dead brother and dead hamster, an idea that sounds absurd now I write it but that seemed about right to me at the time.* To me at least, whether dog or person, grief is grief. It isn't my intention to belittle the loss of a close family member, of course; rather, it's to exalt the status of dogs above that where they currently sit in society.

When Gregory Berns (he of MRI fame) provided strong evidence that the emotional attachment dogs get from us mirrors what we feel for them, it caused something of a Damascene moment in him. A logical conclusion formed in his brain. In 2013, he wrote it down in an opinion piece called 'Dogs are People Too' for *The New York Times*. 'The ability to experience positive emotions, like love and attachment,' he wrote, 'would mean that dogs have a level of sentience comparable to that of a human child. And this ability suggests a rethinking of how we treat dogs.' Berns' central argument was that, if what dog cognition science is telling us is true, then dogs deserve more legal rights than they are currently afforded.

---

* It would turn out I wasn't alone in feeling such mixed emotions. According to Celeste Teo of the National University of Singapore, dog owners regularly face uncertainty about how to grieve for the loss of a companion animal. In her interviews with people mourning the loss of a beloved companion animal, some key themes arise. One is that dog owners feel shame at caring so much. Another is stress, akin to the loss of a human friend or relative. Another is the implicit societal expectations – that, once a companion animal dies, it is time to move on. We probably need to talk about it more.

The argument has a certain polarising quality. On the one hand are those who see this as an inevitable next step in our relationship. That if dogs possess human rights, they can be saved from the worst of what we dish out – the cruelty, the puppy-farms, the abuse. They argue that such legislation might control or even put a stop to the genetic inbreeding commonly practised, even actively pursued, in pedigree dog-breeding circles – activities that see dogs unnecessarily suffering from dermatitis, cataracts, cardiomyopathy, hip and elbow dysplasia, hypothyroidism, back disorders and epilepsy.* But there are those who argue the opposite – that affording dogs the same rights as humans isn't the panacea some imagine it to be. Most notably, some veterinarians are wary of such a ruling. As the law currently stands, pets are considered little more than property so, in cases of accidental malpractice, pet owners cannot claim for more than the price of their animal. The concern is that, with increased rights, come more lawsuits – pet owners suing for more than just cost; suing for suffering caused, work lost, family breakdowns and potentially much more. 'The veterinarians are in a very tricky situation,' argues journalist and author David Grimm. 'When we view our pets like children, we sue like they are children when things go wrong.'

As these and other arguments continue to bubble within civilisation, and within our culture, science will continue to shed light on dogs and their perceptions, emotions, feelings and desires. We should be listening to what this science tells us, but also to the nature of the experiments themselves.

---

* The RSPCA, in particular, are no shrinking violets in the debate. 'When I watch Crufts, what I see in front of me is a parade of mutants,' their then-chief vet Mark Evans once argued on the BBC's *Horizon*. 'Some freakish, garish, beauty pageant that has nothing, frankly, to do with health and welfare.'

What speaks volumes to me, at the end of this journey into the subject, is just how far dogs have come. How wonderful that these loyal animals, who started as little more than laboratory instruments in the late-nineteenth century, have become research collaborators afforded the same rights as human research volunteers today. That Berns and his research colleagues used a consent form, signed by the dogs' owners, that was modelled on a child's consent form. That dogs and their owners could quit the study at any moment. That there were no sedations and no restraints used in these studies. No electric shocks. It was all done through the Brelands' (and Skinner's) great legacy – the triumphant power of positive reinforcement.

It is exciting to have written about such a vibrant and engaging era in the history of animal science. It is humbling to discover that the more scientists gave dogs by way of respect and compassion, the more dogs showed us what they are capable of. In fact, if this book is a love story, then that is its central message: that if we make beasts of animals, we should expect to see beasts reflected back. If we treat them as monsters, we should expect to see them behave like monsters. It is the same with people.

And so, as our world continues to change, as populations rise and climate change puts our future on an uncertain path, I hope we can view other animals in a similar light. For there is nothing that says that dogs will always be our first and only allies in this world. In future, they may be joined by others. I see dogs as laying us a trail. It is up to us to explore it, to understand it, to celebrate it. It is up to us – all of us – to follow and see where next it might lead.

Acknowledgements

Historians, biologists, ethologists, philosophers, psychologists, veterinarians, welfare specialists – I cannot think of a book in which I have had to pull upon so wide a circle of experts. My gratitude to these people is enormous – thank you. Many, if not most, took time to read sample chapters and provided all sorts of guidance and advice. They are, in alphabetical order, Hana Ayoob, Sally Bate, Paulo Clips, Matthew Cobb, Lucy Cooke, Laura Curtis-More, Lavinia Economu, Holly English, Gabi Fleury, Jan Freedman, Jess French, Samadi Galpayage, Brian Hare, Naomi Harvey, Adam Hart, Matilda Holmes, Alexandra Horowitz, Mila Iesanu, Dale Jamieson, Juliane Kaminski, Niki Khan, Robert Kirk, Jaqui Lethaby, Helen Lewis, Adriana Lowe, Rachel Malkani, Philip Martin, Sean McCormack, Ádám Miklósi, Ella Miles, Mary Morrow, Wendy Newton, Joy Pate, Jane Perrone, Chloe Petty, Jim Petty, Lauren Robinson, Daniella Dos Santos, Debbie Stevenson, Alison Skipper, Victoria Stillwell, Iain Strachan, Celeste Teo, Bridgett vonHoldt, Sean Wensley, Abigail Woods, Vanessa Woods. In particular, I'd like to thank Kath Allen and Ruth Kent (my faithful cuttings-sniffer, also of Chan-fame) for time invested on the final draft of this book.

My thanks to The Dogs Trust (www.dogstrust.org.uk), The PDSA (www.pdsa.org.uk) and the RSPCA (www.rspca.org.uk) for their clear guidance on a number of the issues this book covers. And to the Royal Veterinary College, London. Also to the International Association of Anthrozoology (isaz.net) whose annual conferences proved a hotbed of creative ideas, rich debate and, of course, insightful research. And to Animal Aspirations (www.animalaspirations.com) whose aim is to engage students from diverse backgrounds through veterinary science and other animal-related professions.

I am enormously indebted to my agents at UA, Laura Macdougall and Emily Talbot, for all their hard work and support. I am so lucky to have met you – thank you. And to Olivia Davies, also. And to my friends at Bloomsbury who have been so supportive in so many ways for almost (how is this true?) a decade: Jim Martin, Julie Bailey, Angelique Neumann, Lizzy Ewer, Amy Greaves, Sarah Head and Anna (we miss you!) MacDiarmid.

The illustrations that feature in this book come from one of my favourite artists and illustrators, Rebecca Howard. I've always wanted to work with Rebecca and only a small part of the reason is because she's my sister.

On the theme of kin selection, I would also like to offer an enormous thank-you to my mum and dad for all their continued wonderful positivity and support. I really owe you a great deal. I am also so grateful that we shared a dog, Biff. Only in recent years, since having our own dog, have I come to realise what a trial it must have been looking after him while we were all so young.

Lastly, for Emma and Scarlett and Esme. Thank you for allowing me to sit in the corner of our living room, slamming my fingers on a keyboard, without despairing of my happy and selfish passions. Your positivity each day, and your attitude to life, inspire me.

Finally, to Oz – thanks for making the best of all of us.

Research notes and further reading

The opening quotation from this book comes from Dale Jamieson's chapter 'Cognitive Ethology at the End of Neuroscience' in *The cognitive animal: Empirical and theoretical perspectives on animal cognition* by Bekoff, M., Allen, C., & Burghardt, G. M. (2002). Cambridge, Mass: MIT Press. It is used with permission.

Introduction

Without a formal census, dog populations are often something of an estimation, based on veterinary statistics or scaled-up surveys. In this section, I have used figures from the PDSA's annual breakdown of pet numbers in the UK. The figure of 9.5 million dogs for Germany was widely reported in the summer of August 2020, when new laws in Germany saw dog owners legally obliged to give their dogs daily exercise. ('Pets are not cuddly toys – and their needs have to be considered,' said Agriculture Minister Julia Klöckner when the news was announced.) The statistics for dog ownership in the EU come from a 2018 report via Statista, as did the figure of 89.7 million for the USA. Australia's 4.8 million total comes from the 2016 Pet Ownership in Australia report (viewable via animalmedicinesaustralia.org.au). Canada's figure of 7.6 million dogs comes via the research in 2017 by Kynetec (formerly Ipsos) on behalf of the Canadian Animal Health Institute (CAHI).

In this introduction, I mention some excellent guides to what your dog is (and isn't) thinking. Some fantastic authors of books like these include Alexandra Horowitz, John Bradshaw and Marc Bekoff. Their work features across the length and breadth of the book and is referenced appropriately.

Chapter 1

If you are after a really good introduction to Cuvier, Lamarck and Buffon, and how they considered dogs in their theory of how the world came to be, I highly recommend *The Invention of the Modern Dog* by Michael Worboys, Julie-Marie Strange and Neil Pemberton (details below).

Much of the reviews and comments surrounding the release of *On the Origin of Species* mentioned in this chapter come from *Evolutionary Writings: Including the Autobiographies* by Charles Darwin, with a helpful introduction by James A. Secord. An excellent summary of its reception is written by Carolyn Burdett titled 'Darwin and the theory of evolution' and can be found on the British Library website (www.bl.uk). Perhaps the best summary of Morgan, Romanes and Lubbock's influence on early dog science comes from this research article. Direct comments and observations from Lubbock in this chapter come, in part, through his letter in the 'Popular Miscellany' section of *Popular Science*, Vol.25, July 1884.

Catchpole, Caroline (21 February 2013) *'Love and Romance': The Alfred Russel Wallace Correspondence Project*. http://wallaceletters.info/content/love-romance

Darwin, Charles (1859) *On the origin of species by means of natural selection, or, The preservation of favoured races in the struggle for life*. London: J. Murray

— (1871) *The Descent of Man: And Selection in Relation to Sex*. London: J. Murray

— (1872) *The Expression of the Emotions in Man and Animals*. London: J. Murray

Darwin, Charles, and James A. Secord, ed. (2010) *Evolutionary Writings*. Oxford World's Classics

Dawkins, M. S. (October 2006) 'Through animal eyes: What behaviour tells us', *Applied Animal Behaviour Science*, 100 (1–2): 4–10

de Waal, F. (2002) *The ape and the sushi master: cultural reflections by a primatologist*. Penguin

Derr, Mark (2009) 'Darwin's Dogs', *The Bark*. thebark.com/content/darwins-dogs

Eddy, Matthew Daniel (2017) 'The politics of cognition: liberalism and the evolutionary origins of Victorian education', *British Journal for the History of Science*, 50 (4): 677–699

Feuerbacher, Erica & Wynne, C. (2011) 'A History of Dogs as Subjects in North American Experimental Psychological Research', *Comparative Cognition & Behavior Reviews*, 6. 10.3819/ccbr.2011.60001

Galpayage, Samadi & Chittka, Lars (2020) 'Charles H. Turner, pioneer in animal cognition', *Science*, 370. 530–531

Jones, Steve (2000) *Almost Like A Whale: The Origin Of Species Updated*. Black Swan (new edn)

Lovejoy, A. O. (1936) *The great chain of being: A study of the history of an idea*. Cambridge, Mass: Harvard University Press

Mayr, Ernst (1981) *The Growth of Biological Thought*. Cambridge: Harvard, p.330

Parsons, J. H. (1936) 'Conwy Lloyd Morgan. 1852–1936'. Obituary Notices of Fellows of the Royal Society

Romanes, G. (1882) *Animal Intelligence*. D. Appleton and Company

— (1887) 'Experiments on the Sense of Smell in Dogs', *Zoological Journal of the Linnean Society*, Vol.20, Issue 117, June 1887, pp.65–70

Sloan, Phillip (2019) 'Evolutionary Thought Before Darwin', *The Stanford Encyclopedia of Philosophy* (Winter 2019 Edition), Edward N. Zalta (ed.)

Tamatsu, Y., Tsukahara, K., Hotta, M. & Shimada, K. (2007) 'Vestiges of vibrissal capsular muscles exist in the human upper lip', *Clin. Anat.*, 20: 628–631. doi:10.1002/ca.20497

Worboys, Michael, Strange, Julie-Marie and Pemberton, Neil (2018) *The Invention of the Modern Dog: Breed and Blood in Victorian Britain*. Johns Hopkins University Press. *Animals and the Shaping of Modern Medicine: One Health and Its Histories*. Available for free at https://www.palgrave.com

Woods, Abigail, Michael Bresalier, Angela Cassidy, and Rachel Mason Dentinger (2018) *Animals and the Shaping of Modern Medicine: One Health and Its Histories*

Chapter 2

There are, as you can imagine, a few different ways that dog populations have been categorised into ecological groups. In my research, I went with Luigi Boitani's labels of 'owned-restricted', 'owned-unrestricted', 'stray' and 'feral' because it outlines more clearly the distinctions in how dogs have socially bonded with humans. In addition, these categories line up well against those put forward by the World Health Organization in 1988, to allow for better targeting of anti-rabies treatments. These are categorised (by Boitani *et al.*) in Chapter 9 of *The Behavioural Biology of Dogs* (citation below).

For a really superb discussion on the morality of vivisection from a historical perspective, I recommend Robert Kirk's chapter (titled 'The Experimental Animal: In search of a moral ecology of science?') in Kean, H., & Howell, P. (eds) (2018) *The Routledge Companion to Animal–Human History* (1st edn).

The interviews conducted by the Coppingers on how local people viewed street dogs are recounted in Chapter 6 (pp.195–235) of *Genetics and the Behavior of Domestic Animals* (2nd edn), edited by Temple Grandin and Mark J. Deesing (again, the citation is below).

Special thanks to Professor Abigail Woods, animal historian at the University of Lincoln (UK), whose fascinating talk at the International Society for Anthrozoology (ISAZ)'s 2020 conference really got my cogs whirring.

Atickem, A., Bekele, A. & Williams, S. D. (2010) 'Competition between domestic dogs and Ethiopian wolf (Canis simensis) in the Bale Mountains National Park, Ethiopia', *African Journal of Ecology*, 48: 401–407

Bates, A. W. H. (2017) 'Vivisection, Virtue, and the Law in the Nineteenth Century', *Anti-Vivisection and the Profession of Medicine in Britain*. The Palgrave Macmillan Animal Ethics Series. London: Palgrave Macmillan https://doi.org/10.1057/978-1-137-55697-4_2

Beers, D. (2006) *For the Prevention of Cruelty*. Ohio University Press

Bradshaw, J. (2011) *In Defence of Dogs*. London: Allen Lane

Brottman, M. (2014) *The Great Grisby: Two Thousand Years of Exceptional Dogs*. Harper

Butler, J. R. A., Brown, W. Y., du Toit, J. T. (2018) 'Anthropogenic Food Subsidy to a Commensal Carnivore: The Value and Supply of Human Faeces in the Diet of Free-Ranging Dogs', *Animals* (Basel), 2018; 8(5): 67

Cairns, Kylie M., Wilton, & Alan N. (2016) 'New insights on the history of canids in Oceania based on mitochondrial and nuclear data', *Genetica*, 144 (5): 553–565 doi:10.1007/s10709-016-9924-z. PMID 27640201

Coppinger, R; Coppinger, L. (2001) *Dogs: A Startling New Understanding of Canine Origin, Behavior and Evolution*. Prentice Hall

— (2017) *What Is a Dog?* Chicago, IL: University of Chicago Press

Ferguson, D. (2019) 'How the Victorians turned mere beasts into man's best friends', the *Observer*, 19 October 2019

Gorman, James (April 2016) 'The World is Full of Dogs Without Collars', *New York Times*

Grandin, Temple, (ed.), Deesing, Mark, (ed.), and Gale Group (2014) *Genetics and the Behavior of Domestic Animals*. Elsevier Academic Press

Gray, B. (2014) *The Dog in the Dickensian Imagination*. Farnham: Ashgate, p.259

Howell, Philip (2015) *At Home and Astray: The Domestic Dog in Victorian Britain*. University of Virginia Press

Jamieson, P. (1989) 'Animal Welfare: A Movement in Transition', *Law and History: A Collection of Papers Presented at the 1989 Law and History Conference* ed. Suzanne Corcoran. Adelaide: University of Adelaide, p.24

Jenkins, Garry (2011) *A Home Of Their Own*. Transworld Publishing, p.306

Jensen, P. (2007) *The behavioural biology of dogs*. Wallingford, Oxfordshire: CABI International

Kean, H., & Howell, P. (eds) (2018) *The Routledge Companion to Animal–Human History* (1st edn). Routledge

LoPatin-Lummis, N. 'Mad Dogs and Englishmen: Rabies in Britain, 1830–2000'. By Neil Pemberton and Michael Worboys (Houndsmills: Palgrave Macmillan, 2007. 247 pp.), *Journal of Social History*, Vol.43, Issue 3, Spring 2010, pp.752–754

Matthews, Mimi (2018) *The Pug Who Bit Napoleon: Animal Tales of the 18th and 19th Centuries.* Pen and Sword History

Metcalf, Michael F. (1989) 'Regulating slaughter: Animal protection and antisemitism in Scandinavia, 1880–1941', *Patterns of Prejudice*, 23 (3): 32–48

Miklósi, Á. (2015) Ch.8 'Intraspecific social organization in dogs and related forms', *Dog Behaviour, Evolution, and Cognition* (2nd edn). Oxford University Press, pp.172–173

Pal, S., Ghosh, B., & Roy, S. (1998) 'Agonistic behaviour of free-ranging dogs (Canis familiaris) in relation to season, sex and age', *Applied Animal Behaviour Science*, Volume 59, Issue 4, September 1998, pp.331–348

Pemberton N., Worboys, M. (2007) 'Rabies Banished: Muzzling and Its Discontents, 1885–1902', *Mad Dogs and Englishmen: Science, Technology and Medicine in Modern History.* London: Palgrave Macmillan

Striwing, H. (2002) 'Animal Law and Animal Rights on the Move in Sweden', *Animal Law*, 8: 93

Townshend, E. (2009) *Darwin's dogs: How Darwin's pets helped form a world-changing theory of evolution.* London: Frances Lincoln

Trohler, Ulrich; Maehle, Andreas-Holger (1990) 'Anti-vivisection in 19th century Germany and Switzerland: Motives and Methods', in Nicolaas A. Rupke (ed.) *Vivisection in Historical Perspective.* Beckenham, Kent: Croom Helm, Ltd

Williams, Holly (13 August 2010) 'Creature comfort: Why London's first dogs' home was met with howls of derision', the *Independent*

Chapter 3

Should any readers be seeking advice about how to rediscover and celebrate the influence of other forgotten heroes of science like the African-American zoologist Charles H. Turner, here are some suggestions courtesy of the biologist Samadi Galpayage, co-author of the paper that features in Chapter 3, describing Turner's long-overlooked influence on the field of cognition.

'Don't be afraid or discouraged to read very old papers,' Galpayage advises. 'Sometimes these are more difficult to access, as they are not available online; sometimes they feel more difficult to read because of the language (including being written in another language) or even the aesthetics (old prints may not always be pretty, neat and clear), but it's worth trying them out.'

For supervisors reading this, Galpayage recommends encouraging students to share papers or biographies of past and present academics and (importantly) spending time with students, asking questions and discussing the influences that may have been behind the work of the big names in a given field of science. 'Seeing that many experiments done by Turner were later redone and more famous scientists are recognised for them, I would say don't disregard an author just because they're not as well-known as another. The quality and importance of that research can be just as good if not better,' Galpayage tells me. Her final piece of advice: 'Stay curious, be adventurous and feel okay about wandering around sometimes.'

The paper itself is here: Galpayage Dona, H. Samadi & Chittka, Lars. (2020) 'Charles H. Turner, pioneer in animal cognition', *Science*, 370. 530–531.

Onto Pavlov. The names of many of Pavlov's dog were also almost lost to science. Some of the names have only come to light in recent decades, partly thanks to the American neurobiologist Tim Tully who spent eleven years trying to track them down. His journey, published in 2003, culminated in Tully undertaking a pilgrimage to Pavlov's last place of work in Russia where he came upon a photograph album containing the names and profile shots of many of them. Tully immortalised the names of these dogs in the names of fruit-fly strains, mutants of which contain defective genes that code for memory.

The list I quote comes from a postscript in Adams, M., 'The kingdom of dogs: Understanding Pavlov's experiments as human–animal relationships', *Theory & Psychology*, 2020; 30(1): 121–141.

As well as Turner, I feel like Mary Cover Jones has also been somewhat overlooked in the history of psychology and animal science, so allow me to add a few more lines here. As well as undertaking the first experiment in 'unconditioning', Jones went on to lead (alongside her husband) the Oakland Growth Study, the third in a series of important longitudinal studies on children undertaken by Berkeley College. Over a period of years, the study followed the growth of 212 students, up until the end of high school and then, at age 38, age 48 and aged 60. Among the discoveries from this study was the impact that under-age drinking, economic status and age of puberty had on later life. More than a hundred studies were published as a result of this dramatic data-set. It is said that the reason the study worked out was due to the good relations that Jones maintained with her sample population throughout her life. On her death-bed in 1987, her final words were: 'I am still learning about what is important in life.'

For an entertaining journey through Pavlov, Thorndike and Watson I heartily recommend Stephen Budiansky's *If A Lion Could Talk: How Animals Think* – a book I find myself regularly referring back to.

Beck, H. P., Levinson, S. & Irons, G. (2009 Oct) 'Finding Little Albert: a journey to John B. Watson's infant laboratory', *Am Psychol*, 64(7): 605–14 doi: 10.1037/a0017234. PMID: 19824748

Budiansky, S. (1998) *If a Lion Could Talk: Animal Intelligence and the Evolution of Consciousness*. New York: Free Press

Coren, Stanley (2005) *How Dogs Think*. Simon & Schuster

Pavlov, I. P. (1927) *Conditioned Reflexes: An Investigation of the Physiological Activity of the Cerebral Cortex*. Translated and Edited by G. V. Anrep. London: Oxford University Press, p.142

Powell, R. A., & Schmaltz, R. M. (2020) 'Did Little Albert actually acquire a conditioned fear of furry animals? What the film evidence tells us', *History of Psychology*, *24*(2), 164–181

Reiss, B. K. (1990) *A biography of Mary Cover Jones*. Berkeley, CA: Wright Institute

Specter, Michael (24 November 2014) 'Drool: Ivan Pavlov's real quest', *The New Yorker*

Todes, D. P. (2014) *Ivan Pavlov: A Russian life in science*. Oxford University Press

Tully, T. (2003) 'Pavlov's dogs', *Current Biology*, Vol.13, Issue 4, 2003, pp.R117–R119

Watson, John B. (1925) *Behaviorism*. People's Institute Publishing Company

Wood, Jackie (2005) 'The First Nobel Prize for Integrated Systems Physiology: Ivan Petrovich Pavlov, 1904', *Physiology* (Bethesda, Md.), 19. 326–30

Chapter 4

In 1985, a new statue of the brown dog was erected in Battersea Park, funded by the National Anti-Vivisection Society and the British Union for the Abolition of Vivisection. Predictably, University College London and the *British Medical Journal* condemned its unveiling, the latter calling it the 'Libellous Statue At Battersea'. It was removed by officials and put into storage, only to be re-installed in 1994, a little set back from its original position. Commenting about this incident in 1998, the *Nature* writer John Galloway noted: 'It is striking that the monument, or the idea behind it, survived intermittently for nearly 90 years and still had the power to provoke the medical establishment. However, the number of animals used in experiments in the United Kingdom rose from 19,000 each year to more than 3 million in the 90 years following the brown dog's death, a tide the statue did nothing to stem.'

This chapter includes a quote from a recorded conversation with C. S. Sherrington's laboratory technician, E. M. Tansey. You can read this here: http://doi.org/10.1098/rsnr.2007.0037

Galloway, John (13 August 1998) '"Dogged by controversy" – review of Peter Mason's *The Brown Dog Affair*', *Nature*, 394 (6694): 635–636 doi:10.1038/29220. PMID 11645091. S2CID 37893795

Kean, Hilda (1998) *Animal Rights: Political and Social Change in Britain since 1800*. London: Reaktion Books

Lansbury, Coral (1985) *The Old Brown Dog: Women, Workers, and Vivisection in Edwardian England*. Madison: The University of Wisconsin Press

Lind af Hageby, Lizzy; Schartau, Leisa Katherine (1903) *The Shambles of Science: Extracts from the Diary of Two Students of Physiology*. London: E. Bell, OCLC 181077070

Mason, Peter (1997) *The Brown Dog Affair*. London: Two Sevens Publishing

Preece, Rod (2011) *Animal Sensibility and Inclusive Justice in the Age of Bernard Shaw*. Vancouver: UBC

Chapter 5

John Bradshaw's frustration with 'alpha' mentality is eloquently outlined in the marvellous *In Defence of Dogs*, to which this book owes a great debt. I am indebted to a host of experts in the dog behaviour field who made their voices heard to me on this matter, particularly Sean Wensley, Lauren Robinson, Naomi Harvey, Holly English and Gabi Fleury. I'd also like to highlight the great work of Animal Aspirations, a student-led project at the Royal Veterinarian College (London) that aims to encourage diversity in veterinary and animal-related science. Their website can be found at www.animalaspirations.com and they are on Twitter as @AnimalAsp.

Bradshaw, John (2011) *In Defence of Dogs*. Allen Lane

Budiansky, Stephen (2012) *The Truth about Dogs: An Inquiry into the Ancestry, Social Conventions, Mental Habits, and Moral Fiber of Canis Familiaris*. New York: Viking, 2000. Marvin, Garry. Wolf. London: Reaktion

Burch, Mary R. & Bailey, Jon S. (1999) *How Dogs Learn*. New York: Howell Book House

Campbell, Donald T. (1975) 'Reintroducing Konrad Lorenz to Psychology', in Evans, R. I. (ed.), *Konrad Lorenz: The Man and His Ideas*. New York: Harcourt Brace Jovanovich, p.106

Cohen, S. (2010) 'Canine Dominance: Is the Concept of the Alpha Dog Valid?', *Psychology Today*

Coren, Stanley (1995) *The Intelligence of Dogs: A Guide To The Thoughts, Emotions, And Inner Lives Of Our Canine Companions*. New York: Bantam Books

Csânyi, Vilmos (2005) *If Dogs Could Talk: Exploring the Canine Mind* (1st American edn). New York: North Point

Druzhkova, A. S., Thalmann, O., Trifonov, V. A., Leonard, J. A., Vorobieva, N. V., Ovodov, N. D., *et al.* (2013) 'Ancient DNA Analysis Affirms the Canid from Altai as a Primitive Dog', *PLoS ONE*, 8(3): e57754 https://doi.org/10.1371/journal.pone.0057754

Eyman, Scott (2005) *Lion of Hollywood: The Life and Legend of Louis B. Mayer*. Robson

Gray, J. (2018) 'The History of Dog Training, Influential Movement Creators in the Industry and the Impact of Training and Behaviour Adjustment', *Academic Journal of Canine Science*, 27 November 2018

Hare, Brain & Vanessa Woods (2013) *The Genius of Dogs*. New York: Dutton

Herzog, Hal (2011) *Some We Love, Some We Hate, Some We Eat: Why It's so Hard to Think Straight about Animals*. New York, NY: Harper Perennial

Horowitz, Alexandra (2009) *Inside of a Dog: What Dogs See, Smell, and Know* (1st Scribner Hardcover edn). New York: Scribner

Kaminski, J., Waller, B., Diogo, R., Hartstone-Rose, A. & Burrows, A. (2019) 'Evolution of facial muscle anatomy in dogs', *Proceedings of the National Academy of Sciences*, July 2019, 116 (29) 14677–14681

Loog, L., Thalmann, O., Sinding, M.-H. S., *et al.* (2020) 'Ancient DNA suggests modern wolves trace their origin to a Late Pleistocene expansion from Beringia', *Molecular Ecology*, 2020; 29: 1596–1610

Lorenz, Konrad (1953) *Man Meets Dog* (Marjorie Kerr Wilson, Trans.) Hagerstown, MA: Kodansha America, 1994

— (1952) *King Solomon's Ring; New Light on Animal Ways.* New York: Crowell

Mech, L. D. (1970) *The Wolf: The Ecology and Behavior of an Endangered Species.* Natural History Press (Doubleday Publishing Co., N.Y.) p.389 (Reprinted in paperback by University of Minnesota Press, May 1981)

Most, K. (1954) *Training Dogs* (J. Cleugh, Trans.) New York: Dogwise Publishing

Pryor, Karen (1984) *Don't Shoot the Dog: The New Art of Teaching and Training.* New York: Bantam Books

Saunders, Blanche (1969) *Training You to Train Your Dog.* New York: Howell Book House

Sax, Boria (1997) 'What is a "Jewish Dog"? Konrad Lorenz and the Cult of Wildness', *Society and Animals*, 5 (1): 3–21

Serpell, J. (ed.) (2016) *The Domestic Dog: Its Evolution, Behavior and Interactions with People* (2nd edn). Cambridge: Cambridge University Press

Twain, M., & Kiskis, M. J. (2010) *Mark Twain's own autobiography: The chapters from the North American review.* Madison, Wis: University of Wisconsin Press

Zimen, Erik (1981) *The Wolf : His Place in the Natural World.* London: Souvenir

Chapter 6

For a more detailed account of Marian Breland, I highly recommend Cook-Hasley & Wiebers (1999) *Marian Breland Bailey: A Pioneer in the History of Applied Animal Psychology.* Henderson State University.

I should add somewhere that this chapter (actually, many chapters of this book) owe a debt to Wynne and Feuerbacher's paper entitled 'A history of dogs as subjects in North American experimental psychological research'. If you are enjoying this book, this historical paper is well worth a read and is thankfully very easy to digest. (The details are below.) Incidentally, I agree fully with the assertion that Wynne and

Feuerbacher put forward in the paper, specifically that a 'rediscovery of this literature can only aid research being conducted today, including rejuvenating old questions, suggesting new ones, and highlighting useful methods for current issues.'

Cook-Hasley & Wiebers (1999) *Marian Breland Bailey: A Pioneer in the History of Applied Animal Psychology.* Henderson State University https://www.hsu.edu/uploads/pages/199-0afmarian_breland_bailey.pdf

Feuerbacher, E. & Wynne, C. (2011) 'A History of Dogs as Subjects in North American Experimental Psychological Research', *Comparative Cognition & Behavior Reviews*, 6: 46–71

Fishkoff, S. (4 November 1999) 'Pecking order: Whatever happened to the chickens who worked the tic-tac-toe game on Cannery Row?', *Montery County Weekly*

Gillaspy, J. A., & Bihm, E. M. (July 16, 2007) *Keller Bramwell Breland (1915–1965).* The Encyclopedia of Arkansas History & Culture

Haggbloom, Steven J., Warnick, Renee, Warnick, Jason E., Jones, Vinessa K., *et al.* (2002) 'The 100 most eminent psychologists of the 20th century', *Review of General Psychology*, 6 (2): 139–52

Horowitz, Alexandra (2009) *Inside of a Dog: What Dogs See, Smell, and Know* (1st Scribner Hardcover edn). New York: Scribner

Skinner, B. F. (1972) *Beyond freedom and dignity.* New York: Vintage Books

— (1938) *The Behavior of Organisms.* New York: Appleton-Century-Crofts

Timberlake, W., Lucas, G. A. (November 1, 1985) 'The basis of superstitious behavior: chance contingency, stimulus substitution, or appetitive behavior?', *J Exp Anal Behav*, 44 (3): 279–299

Chapter 7

Unitree's A1 robot was released in 2020 as a rival to the robot dogs made by Boston Dynamics, the robotics company made famous across social media (and *that* episode of Netflix's *Black*

Mirror). There are plenty of videos of A1 online, but its release into the world was covered by *Business Insider* on 2 May 2020. ('There's a new competitor in the world of eerily lifelike four-legged robots – take a closer look at Unitree's robot dogs', https://tinyurl.com/yxaox9n2.)

To keep this chapter from becoming a bit too sprawling, I have had to pull back a number of the key figures in the development of the 'Computer as Brain / Brain as Computer' hypothesis. A full and highly readable account is expressed in glorious detail by the author and zoologist, Matthew Cobb in *The Idea of the Brain* (citation below).

Arbib, Michael (2003) 'Rana computatrix to human language: Towards a computational neuroethology of language evolution', *Philosophical transactions. Series A, Mathematical, physical, and engineering sciences*, 361. 2345–79

Barandiaran, Xabier & Chemero, Anthony (2009) 'Animats in the Modeling Ecosystem', *Adaptive Behavior*, 17. 287–292 10.1177/1059712309340847

Boden, Margaret A. (2006) *Mind as Machine: A History of Cognitive Science*. Oxford: Clarendon Press, p.227

Breland, K. & Breland, M. (1961) 'The Misbehavior Of Organisms', Animal Behavior Enterprises, Hot Springs, Arkansas. First published in *American Psychologist*, 16, 681–684

Bynum, Terrell (2002) 'Norbert Wiener's Vision: The Impact of "the Automatic Age" on Our Moral Lives' https://tinyurl.com/knhfpk26

Cobb, Matthew (2020) *The Idea of the Brain: A History*. London: Profile Books Ltd

Cowan, N., Morey, C. C., Chen, Z. (2007) *The legend of the magical number seven*. In Sergio Della

Sala (ed.) *Tall tales About the Brain: Separating Fact from Fiction*. Oxford University Press

Gray, Paul (29 March 1999) 'Alan Turing – Time 100 People of the Century', *Time*

Heunis, Christoff & van den Heever, Dawie (2014) *Design and Construction of an Autonomous Cricket Robot to Demonstrate the Language of Neurons*, 10.13140/RG.2.1.3593.2564

Jeffery, K. (25 January 2017) *Maps in the head.* Aeon
Miller, G. A. (1956) 'The magical number seven, plus or minus two: some limits on our capacity for processing information', *Psychological Review*, 63(2), 81–97 https://doi.org/10.1037/h0043158
Nicole Chapuis & Christian Varlet (1987) 'Short cuts by dogs in natural surroundings', *Quarterly Journal of Experimental Psychology*, Section B, 39:1, 49–64
O'Keefe, J. & Nadel, L. (1978) *The Hippocampus as a Cognitive Map.* Oxford University Press
Turing, Alan (1950) 'Computing Machinery and Intelligence', *Mind*, 49 (236): 433–460
Wilson, S. W. 'The animat path to AI', in J.-A. Meyer and S. Wilson (eds) (1991) *From Animals to Animats*, pp.15–21. Cambridge, MA: MIT Press http://www.eskimo.com/~wilson/ps/animat.pdf

Chapter 8

I am indebted to Joy Pate of Penn State University for her assistance with an early draft of this chapter. Details of her research into the development rate of puppies, largely through 'fieldwork' undertaken by co-author Mary Morrow, are included in the references below.

In the second half of this chapter, we hear of Seligman's first labelling of the phenomena of learned helplessness in dogs – a finding that would later inspire his influence on Cognitive Behavioural Therapy. To spare readers too many gruesome details in the main text, I have decided to move the details of Seligman's experiments involving dogs to this section for readers to digest in more detail, should they wish.

In the first stage of the experiment, Seligman's dogs weren't yet placed in shuttle-boxes at all. Instead, they were harnessed away from the shuttle-box. The first group of dogs – the control group – were simply held still in temporary harnesses.

In the second group, the harnessed dogs were provided with a lever that could be used to put an end to a series of electric shocks induced in them by an unseen human experimenter. (Naturally, to end their suffering, the dogs soon became conditioned to pulling this lever.) The experimental set-up for the second group faced was undoubtedly very hard on the subjects, but the dogs in the third group had it the worst. These dogs were paired up with dogs from the second group of dogs – the ones that could work the lever. If one dog received an electric shock, so would the other. The dogs in group three had a lever too, except the levers of the dogs in group three did not work. To the dogs in group three, strapped up to the group two counterparts, the electric shocks were apparently random and, crucially, uncontrollable. They had only one choice: to endure. These dogs were crucial to the experiment. The trauma experienced by these dogs would shape how they responded to the second stage of Seligman and Maier's experiments.

In the second stage of the experiment, the shuttle-boxes were brought out. Predictably, when put into the shuttle-boxes, the first and second groups of dogs quickly learned that, if shocked, a quick vault to the other side of the shuttle-box would make the shocks cease. The third group, however, did not. Emotionally damaged from the first phase of the experiment, these dogs did nothing. They simply remained still where they were and took the shocks. They whined, sure. They winced. They endured. For these group three dogs, the situation was not deemed one capable of their control, so they settled for their fate. Seligman and colleagues coined the term that would describe their emotional state: it was one of 'learned helplessness'. In all, about two-thirds of the animals in this third group exhibited this phenomenon. In this state, subjects showed increased signs of stress; they had reduced appetite for food (and sex), and they suffered lethargy.

They were, the researchers agreed, displaying learned helplessness.

It brings me no pleasure to relay these facts, but sometimes looking back from the mountain of experience helps steel our resolve to never slip backwards.

As well as Seligman, this chapter also alludes at the end to the growing anti-vivisectionist movement in the USA. Much of this comes from the national outcry in 1965 caused by a stolen dog named Pepper, later found to have been sold by a third party to a research laboratory. Pepper was later discovered to have died while undergoing surgery involving a cardiac pacemaker.

According to Daniel Engber, who in 2009 authored a series of essays on the case for *Slate* and Hal Herzog (author of the excellent *Some we love, Some we hate, Some we eat – Why it's so hard to think straight about animals*), Pepper's story signalled a landmark change in American pet politics. It created a media sensation, built upon the public's widespread fear that their pets were fair game to be stolen and used in medical research. Following a *Life* magazine article on the issue at the time ('Concentration Camp for Dogs'), government officials received more angry letters about the issue than during the entire Vietnam War. The unrelenting pressure eventually saw lawmakers speed up the passing of the Animal Welfare Act of 1966. Pepper did not die in vain.

Battaglia, C. L. (2009) 'Periods of early development and the effects of stimulation and social experiences in the canine', *J Vet Behav*, 2009; 4: 203–210

Coren, Stanley (2005) *How Dogs Think*. Simon & Schuster

Dewsbury D. A. (ed.) (1985) *Studying Animal Behavior: Autobiographies of the Founders*. University of Chicago Press, Chicago. (Original edn, Lewisburg, Bucknell University Press)

Dewsbury, Donald (2011) 'A History of the Behavior Program at the Jackson Laboratory: An Overview', *Journal of comparative psychology* (Washington, D.C.: 1983), 126. 31–44 10.1037/a0021376

Dietza, L. *et al.* (2018) 'The importance of early life experiences for the development of behavioural disorders in domestic dogs', *Behaviour*, 155. 83–114

Engber, Daniel (1 June 2009) 'Pepper: the stolen dog that changed American science', Slate.com. Retrieved 23 May 2012 (first of a five-part article analysing the impact of Pepper)

Herzog, Hal *Some we love, Some we hate, Some we eat – Why it's so hard to think straight about animals.* Harper Perennial, pp.223–224

Konnikova, M. (January 14, 2015) 'Trying to Cure Depression, but Inspiring Torture', *The New Yorker*

Maier, Steven F., & Seligman, Martin E. P. (2016) 'Learned helplessness at fifty: Insights from neuroscience', *Psychological review*, Vol.123,4: 349–67

Matthew, H. C. G., Harrison, B. (eds) (2004-09-23) *The Oxford Dictionary of National Biography.* Oxford University Press

Morrow, M., Ottobre, J., Ottobre, A., Neville, P., St-Pierre, N., Dreschel, N., & Pate, J. L. (2015) 'Breed-dependent differences in the onset of fear-related avoidance behavior in puppies', *Journal of Veterinary Behavior: Clinical Applications and Research*, 10(4), 286–294

Scott, John Paul (1986) 'John Paul Scott Oral History'. Oral History Collection 3 https://mouseion.jax.org/oral_history

Scott, J. P., Fuller, J. L. (1965) *Genetics and the Social Behavior of the Dog.* University of Chicago Press

Seligman, M. E., Maier, S. F., & Geer, J. H. (1968) 'Alleviation of learned helplessness in the dog', *Journal of Abnormal Psychology*, 73(3, Pt1), 256–262

Valsecchi, P., Previde, E. P., Accorsi, P. A., & Fallani, G. (2010) 'Development of the attachment bond in guide dogs', *Applied Animal Behaviour Science*, 123, 43–50

The illustration that heads up this chapter is based on a photo given by John Paul Scott to Donald Allen Dewsbury, the American comparative psychologist. https://creativecommons.org/licenses/by/4.0/

Chapter 9

Of the books I still have from my university days, two books are so well thumbed that they have become little more than two cuboids of loose paper. I found myself referring to these books a great deal during the formulation of this chapter. They are Stephen Budiansky's *If A Lion Could Talk* (1998) and Marc Hauser's *Wild Minds* (2001). If you are interested in the story of both Nagel and Gallup's involvement in the animal cognition scene, these books are worthy of a revisit and have stood the test of time mightily.

In the latter stages of this chapter, there is mention of Irene Pepperberg and her incredible research on African Grey Parrots. There is a delightful account of Pepperberg and her research in Virginia Morrell's utterly brilliant *Animal Wise* – another well-thumbed book on my shelf. In 2012, Pepperberg would be among those to sign the Cambridge Declaration on Consciousness, in which a prominent group of cognitive neuroscientists, neurophysiologists, neuroanatomists gathered at the University of Cambridge to set out their stall. Here is what it said:

'Convergent evidence indicates that non-human animals have the neuroanatomical, neurochemical, and neurophysiological substrates of conscious states along with the capacity to exhibit intentional behaviors.'

'Consequently, the weight of evidence indicates that humans are not unique in possessing the neurological substrates that generate consciousness.'

Allen, Colin & Trestman, Michael (2020) *Animal Consciousness, The Stanford Encyclopedia of Philosophy*. Edward N. Zalta (ed.), Winter 2020 Edition

Bálint, A., Andics, A., Gácsi, M. *et al.* (2020) 'Dogs can sense weak thermal radiation', *Sci Rep*, 10, 3736

Bekoff, Mark (2018) *Canine Confidential: Why Dogs Do What They Do*. University of Chicago Press

Budiansky, Stephen (1998) *If a Lion Could Talk: Animal Intelligence and the Evolution of Consciousness*. The Free Press

Cazzolla Gatti, Roberto (2015) 'Self-consciousness: beyond the looking-glass and what dogs found there', *Ethology Ecology & Evolution*, 1–9

Coren, Stanley (7 July 2011) 'Does My Dog Recognize Himself in a Mirror?', *Psychology Today*

Darwin, Charles (1838) *Notebook*, Lines 79 & 196–197

— (1871) *The Descent of Man: And Selection in Relation to Sex*. London: J. Murray

Dawkins, R. (1989) *The Selfish Gene*. Oxford: Oxford University Press

Dennett, Daniel C. (1991) *Consciousness Explained*. Boston: Little, Brown and Company

Drumm, Patrick & Ovre, Christopher (April 2011) 'A batman to the rescue', *Monitor on Psychology*, 42 (4): 24

Gallup, G. G. J., Burch, R. L. & Platek, S. M. (2002) 'Does Semen Have Antidepressant Properties?', *Arch Sex Behav*, 31, 289–293

Goodier, J. (2003) 'Animals and Science: A Guide to the Debates', *Reference Reviews*, Vol.17 No.4, pp.38–39 https://doi.org/10.1108/09504120310473560

Griffin, Donald & Galambos, Robert (April 1941) 'The Sensory Basis of Obstacle Avoidance by Flying Bats', *Journal of Experimental Zoology*, 86 (3): 481–505

Jeffreys, Derek S. (2002) 'Review of Donald R. Griffin. 2001. Animal Minds: Beyond Cognition to Consciousness', *American Journal of Bioethics*, 2:4, 70–71

Marta Zaraska (11 October 2017) 'The Sense of Smell in Humans is More Powerful Than We Think', *Discover* magazine

McGreal, Scott A. (12 September 2012) 'Semen an Antidepressant? Think Again', *Psychology Today*

Nagel, Thomas (1974) 'What Is It Like to Be a Bat?', *Philosophical Review*, 83 (4): 435–450

Pepperberg, Irene M. (1 April 2006) 'In Memoriam: Donald R. Griffin, 1915–2003', *The Auk*, 123(2), 595–597

Steen, J. & Mohus, I. & Kvesetberg, T. & Walløe, Lars. (1996) 'Olfaction in bird dogs during hunting', *Acta physiologica Scandinavica*, 157. 115–9

Suarez, Susan D. & Gallup, Gordon G. (1981) 'Self-recognition in chimpanzees and orangutans, but not gorillas', *Journal of Human Evolution*, Vol.10, Issue 2, pp.175–188

Yong, Ed (13 February 2017) 'What Mirrors Tell Us About Animal Minds', *The Atlantic*

Zimmer, Carl (21 April 2015) 'When Darwin Met Another Ape', *National Geographic*

Chapter 10

A brief word on dog 'citizen science': Today, Miklósi, Hare and other leading dog cognition experts work together as senior advisors on a citizen-science project called Dognition that is attempting to obtain knowledge about the cognitive hardware of dogs on a much wider scale. The hands-on project encourages owners and their dogs to take part in a series of fun activities, each of which provides insights into the suite of cognitive skills that family dogs possess. The reward for Hare and colleagues is that, through Dognition, they can acquire widespread data about dogs, plotting the results of their cognition tests against age, location or breed, for instance. The reward for owners is three-fold. First, that they can get a better feel for their dogs. Second, a warm glow in the knowledge that their dog is contributing to research into dog brains. Third, the dogs get rewards – namely, a cute little reward badge that says, according to the results, whether their dog is a 'charmer' ('Canis Irresistabilis'), a 'socialite' ('Friend To All'), a 'Renaissance dog' ('A Dog Of All Trades') or one of another tongue-in-cheek categories. The results of Dognition, and studies like it, will keep scientists busy for decades. To find out more or to enrol your dog go to www.dognition.com.

This chapter in particular owes an enormous amount to Virginia Morell's brilliant *Animal Wise* – it was through this book's chapter on dogs that I first got a feel for the important impact that the Family Dog Project had played in the recent history of dogs in science. Morell is nothing short of a giant in science-writing and I owe her a great debt. Her regular contributions to *Science* can be found here: www.sciencemag.org/author/virginia-morell.

The chapter also delves into great apes and the degree to which they exhibit a theory of mind that may be like our own. Krupenye's article on the issue ('Great apes anticipate that other individuals will act according to false beliefs') is a really interesting read, as is Frans De Waal's opinion piece in *Science* ('Apes know what others believe'). As he so often does, Ed Yong (of *The Atlantic*) offers a delicious commentary on the subject in his article 'Apes Might Know That You Don't Know What They Know'. Details of each article are included below.

Claudia Kawczynska (November 2008) 'Canine Intelligence: Understand Dogs' Minds', *The Bark*. https://thebark.com/content/canine-intelligence-understand-dogs-minds

Abdai, J., Ferdinandy, B., Lengyel, A., Miklósi, Á. (2020) 'Animacy perception in dogs (Canis familiaris) and humans (Homo sapiens): Comparison may be perturbed by inherent differences in looking patterns', *Journal of Comparative Psychology*, doi: 10.1037/com0000250

Bálint, A., Andics, A., Gácsi, M., Gábor, A., Czeibert, K., Luce, M. L., Miklósi, Á., Kröger, R. H. H. (2020) 'Dogs can sense weak thermal radiation', *Scientific Reports*, 10: 3736 doi: 10.1038/s41598-020-60439-y

Bódizs, R., Kis, A., Gácsi, M., Topál, J. (2020) 'Sleep in the dog: comparative, behavioral and translational relevance', *Current Opinion in Behavioral Sciences*, 33: 25–33

Bognár, Zs., Piotti, P., Szabó, D., Le Nézet, L., Kubinyi, E. (2020) 'A novel behavioural approach to assess responsiveness to auditory and visual stimuli before cognitive testing in family dogs', *Applied Animal Behaviour Science*, in press. doi: 10.1016/j.applanim.2020.105016

Bradshaw, John (2017) *The Animals Among Us: The New Science of Anthrozoology*. Penguin Books Limited

Bunford, N., Hernández-Pérez, R., Farkas, E., Cuaya, L., Szabó, D., Szabó, Á., Gácsi, M., Miklósi, Á., Andics, A. (2020) 'Comparative brain imaging reveals analogous and divergent patterns of species- and face-sensitivity in humans and dogs',

Journal of Neuroscience, JN-RM-2800-19; doi: 10.1523/JNEUROSCI.2800-19.2020

Burkeman, Oliver (21 January 2015) 'Why can't the world's greatest minds solve the mystery of consciousness?', the *Guardian*

Call, J., Tomasello, M. (2008) 'Does the chimpanzee have a theory of mind? 30 years later', *Trends Cogn. Sci*, 12, 187–192

Carballo, F., Cavalli, C. M., Gácsi, M., Miklósi, Á. & Kubinyi, E. (2020) 'Assistance and therapy dogs are better problem solvers than both trained and untrained family dogs', *Frontiers in Veterinary Science*, 7: 164

Csányi, Vilmos (2005) *If Dogs Could Talk: Exploring the Canine Mind* (1st American edn). New York: North Point

Czeibert, K., Sommese, A., Petneházy, Ö., Csörgó, T., Kubinyi, E. (2020) 'Digital endocasting in comparative canine brain morphology', *Frontiers in Veterinary Sciences*, 7: 565315

de Waal, Frans B. M. (2016) 'Apes know what others believe', *Science*, 7 October 2016, Vol.354, Issue 6308, pp.39–40

Delanoeije, J., Gerencsér, L., Miklósi, Á. (2020) 'Do dogs mind the dots? Investigating domestic dogs' (Canis familiaris) preferential looking at human-shaped point-light figures'. Wiley Online Library, 00: 1–14

Feuerbacher, E. N., & Wynne, C. D. L. (2011) 'A history of dogs as subjects in North American experimental psychological research', *Comparative Cognition & Behavior Reviews*, 6, 46–71 https://doi.org/10.3819/ccbr.2011.60001

Fugazza, C., Miklósi, Á. (2020) 'Depths and limits of spontaneous categorization in a family dog', *Scientific Reports*, 10: 3082

— (2020) 'How to make smoke without fire. Minds are (not) just trainable machines', *Learning & Behavior*, pressed online. doi: 10.3758/s13420-020-00447-0

Fugazza, C., Pongrácz, P., Pogány, Á., Lenkei, R., Miklósi, Á. (2020) 'Mental representation and episodic-like memory of own actions in dogs', *Scientific Reports*, 10: 10449

Gábor, A., Gácsi, M., Szabó, D., Miklósi, Á., Kubinyi, E., Andics, A. (2020) 'Multilevel fMRI adaptation for spoken word processing in the awake dog brain', *Scientific Reports*, 10: 11968

Gergely, A., Kiss, O., Reicher, V., Iotchev, I., Kovács, E., Gombos, F., Benczúr, A., Galambos, Á., Topál, J., Kis, A. (2020) 'Reliability of family dogs' sleep structure scoring based on manual and automated sleep stage identification', *Animals*, 10: 0

Gnanadesikan, G. E., Hare, B., Snyder-Mackler, N., Call, J., Kaminski, J., Miklósi, Á., MacLean, E. L. (2020) 'Breed differences in dog cognition associated with brain-expressed genes and neurological functions', *Integrative & Comparative Biology*, icaa112

Goldman, Jason (2013) 'Can this sneaky chimp read minds?', BBC Future. www.bbc.com/future/article/20131125-can-this-sneaky-chimp-read-minds

Gunde, E., Czeibert, K., Gábor, A., Szabó, D., Kis, A., Arany-Tóth, A., Andics, A., Gácsi, M., Kubinyi, E. (2020) 'Longitudinal volumetric assessment of ventricular enlargement in pet dogs trained for functional magnetic resonance imaging (fMRI) studies', *Veterinary Sciences*, 7: 127

Hare, B., & Tomasello, M. (1999) 'Domestic dogs (Canis familiaris) use human and conspecific social cues to locate hidden food', *Journal of Comparative Psychology*, 113(2), 173–177 https://doi.org/10.1037/0735-7036.113.2.173

Hare, B., Call, J., Tomasello, M. (2001) 'Do chimpanzees know what conspecifics know?', *Anim. Behav.* 61, 139–151

Iotchev, B. I., Reicher, V., Kovács, E., Kovács, T., Kis, A., Gácsi, M., Kubinyi, E. (2020) 'Averaging sleep spindle occurrence in dogs predicts learning performance better than single measures', *Scientific Reports*, 10: 22461

Jónás, D., Sándor, S., Tátrai K., Egyed, B., Kubinyi, E. (2020) 'A preliminary study to investigate the genetic background of longevity based on whole-genome sequence data of two Methuselah dogs', *Frontiers in Genetics*, 11: 315

Kano, F., Call, J. (2014) 'Great apes generate goal-based action predictions: An eye-tracking study', *Psychol. Sci.* 25, 1691–1698

Krupenye, C., Kano, F., Hirata, S., Call, J., & Tomasello, M. (2016) 'Great apes anticipate that other individuals will act according to false beliefs', *Science*, 354(6308), 110–113

Kubinyi, E., Iotchev, I. B. (2020) 'A preliminary study toward a rapid assessment of age-related behavioral differences in family dogs', *Animals*, 10: 1222

Lehner, L., Czeibert, K., Benczik, J., Jakab, C., Nagy, G. (2020) 'Transcallosal removal of a choroid plexus tumor from the lateral ventricle in a dog', Case report, *Frontiers in Veterinary Sciences*, 7: 536

Lehner, L., Czeibert, K., Nagy, G. (2020) 'Two different indications of ventriculoperitoneal and cystoperitoneal shunting in six dogs', *Acta Veterinaria Hungarica*, 68: 95–104 doi: 10.1556/004.2020.00010

Lenkei, R., Újváry, D., Bakos, V., Faragó, T. (2020) 'Adult, intensively socialized wolves show features of attachment behaviour to their handler', *Scientific Reports*, 10: 17296

Lindsay, Steven R. (2008) *Handbook of Applied Dog Behavior and Training*. Ames, Iowa, USA: Iowa State UP

Magyari, L., Huszár, Zs., Turzó, A., Andics, A. (2020) 'Event-related potentials reveal limited readiness to access phonetic details during word processing in dogs', *Royal Society Open Science*, 7: 200851

Morell, Virginia (28 August 2009) 'Going to the Dogs', *Science*, Vol.325, Issue 5944, p.1054

— (2013) *Animal Wise: How We Know Animals Think and Feel*. New York: Broadway Books

Pérez Fraga, P. P., Gerencsér, L., Lovas, M., Újváry D., Andics, A. (2020) 'Who turns to the human? Companion pigs' and dogs' behaviour in the unsolvable task paradigm', *Animal Cognition*, press online

Pérez Fraga, P., Gerencsér, L., Andics, A. (2020) 'Human proximity seeking in family pigs and dogs', *Scientific Reports*, 10: 20883

Premack, D. & Woodruff, G. (1978) 'Does the chimpanzee have a theory of mind?', *Behav. Brain Sci.* 1, 515–526

Pughe, D. L. (November 2008, Updated February 2015) 'Studying the dog: A friendly pack is scaling ivory towers on campuses worldwide', *The Bark*. https://thebark.com/content/studying-dog

Reicher, V., Kis, A., Simor, P., Bódizs, R., Gombos, F., Gácsi, M. (2020) 'Repeated afternoon sleep recordings indicate first-night-effect-like adaptation process in family dogs', *Journal of Sleep Research*, 00: e12998

Salamon, A., Száraz J., Miklósi, Á., Gácsi, M. (2020) 'Movement and vocal intonation together evoke social referencing in companion dogs when confronted with a suspicious stranger', *Animal Cognition*, 23, 913–924 (2020)

Szabó, D., Gábor, A., Gácsi, M., Faragó, T., Kubinyi, E., Miklósi, Á., & Andics, A. (2020) 'On the face of it: no differential sensitivity to internal facial features in the dog brain', *Frontiers in Behavioral Neuroscience*, 14: 25

Topál, József & Miklósi, Ádám & Csányi, Vilmos & Antal, Dóka (1998) 'Attachment Behavior in Dogs (Canis familiaris): A New Application of Ainsworth's (1969) Strange Situation Test', *Journal of comparative psychology* (Washington, D.C.: 1983), 112. 219–29

Torda, O. J., Vékony, K., Junó, V. K., Pongrácz, P. (2020) 'Factors affecting Canine obesity seem to be independent of the economic status of the country – A survey on Hungarian Companion Dogs', *Animals*, 10: 1267

Turcsán, B., Tátrai, K., Petró, E., Topál, J., Balogh, L., Egyed, B., Kubinyi, E. (2020) 'Comparison of behavior and genetic structure in populations of family and kenneled beagles', *Frontiers in Veterinary Science*, 7: 183

Turcsán, B., Wallis, L., Berczik, J., Range, F., Kubinyi, E., Virányi, Zs. (2020) 'Individual and group level personality change across the lifespan in dogs', *Scientific Reports*, 10: 17276

Wallis L. J., Iotchev I., Kubinyi, E. (2020) 'Assertive, trainable and older dogs are perceived as more dominant in multi-dog households', *PLoS One*, 15: e0227253

Wallis, L., Szabo, D., Kubinyi, E. (2020) 'Cross-sectional age differences in canine personality traits; influence of breed, sex, previous trauma, and dog obedience tasks', *Frontiers in Veterinary Science*, 6: 493 doi: 10.3389/fvets.2019.00493

Watowich, M. M., MacLean, E. L., Hare. B., Call, J., Kaminski, J., Miklósi, Á., Synder-Mackler, N. (2020) 'Age influences domestic dog cognitive performance independent of average breed lifespan', *Animal Cognition*, (): 1–11

Yong, E. (15 November 2019) 'Apes Might Know That You Don't Know What They Know', *The Atlantic* www.theatlantic.com /science/archive/2019/11/do-apes-have-theory-mind /602038/

Chapter 11

This chapter in particular owes a debt to Alexandra Horowitz, author of *Inside of a Dog* and *Our dogs, ourselves*. If you are interested in dogs and their capacity for social cognition, Alexandra's work has it all: research, experience and plenty of personal anecdotes and observations. She is also a fine – one of the finest of – animal writers. Her website is https://alexandrahorowitz.net/

I am also grateful to Juliane Kaminski (herself a legend in the field of animal cognition!) for her recollections of Rico. You can see footage of Rico fast-mapping on the following YouTube page: https://tinyurl.com/fhdhnjbx

Lyudmila Trut's comments about silver foxes come from the BBC's *Horizon*. The episode titled 'The Secret Life of the Dog' aired on BBC Two in 2010.

Rollovers and some of the other well-known play behaviours noted in this chapter are discussed in this blog-post for *Scientific American*, written by Julie Hecht on 9 January 2015, 'Why Do Dogs Roll Over During Play?' https://blogs.scientificamerican.com/dog-spies/why-do-dogs-roll-over-during-play/

Bekoff, M and Byers, J. (eds) (1998) *Animal Play: Evolutionary, Comparative and Ecological Perspectives*. Cambridge: Cambridge University Press

Bekoff, M. (1975) 'The communication of play intention: Are play signals functional?', *Semiotica*, 15(3), 231–240

Bekoff, M. (2018) 'Animal Studies Repository' 4-1984, *Social Play Behavior*

Bloom, Paul (2004) 'Can a dog learn a word?', *Science* (New York, N.Y.), 304. 1605–6

Bradshaw, John W. S., Pullen, Anne J., Rooney, Nicola J. 'Why do adult dogs "play"?', *Behavioural Processes. New Directions in Canine Behavior.* Edited by Monique Udell. Vol.110, pp.82–87

Burghardt, G. M. (2014) 'A Brief Glimpse at the Long Evolutionary History of Play', *Animal Behavior and Cognition*, 1(2), 90–98

Hansen Wheat C., Fitzpatrick J. L., Rogell B., Temrin H. (2019) 'Behavioural correlations of the domestication syndrome are decoupled in modern dog breeds', *Nat Commun*. 2019; 10(1): 2422

Hare B., Plyusnina I., Ignacio N., Schepina O., Stepika A., Wrangham R., Trut L. (2005) 'Social cognitive evolution in captive foxes is a correlated by-product of experimental domestication', *Current Biology* Vol.15, Issue 3, pp.226–230

Horowitz Alexandra (2008) 'Attention to attention in domestic dog (Canis familiaris) dyadic play', *Animal Cognition*, 12 (1) 107–118

— (2009) *Inside of a Dog: What Dogs See, Smell, and Know* (1st Scribner Hardcover edn). New York: Scribner

Kaminski, J., Pitsch, A., Tomasello, M. (2012) 'Dogs steal in the dark', *Animal Cognition*, 2013, Vol.16(3), pp.385–394

Kaminski, Juliane & Call, Josep & Fischer, Julia (2004) 'Word Learning in a Domestic Dog: Evidence for "Fast Mapping"', *Science* (New York, N.Y.), 304. 1682–3

— (2008) 'Prospective object search in dogs: mixed evidence for knowledge of What and Where', *Animal Cognition*, 11, 367–371

Kukekova, Anna & Temnykh, Svetlana & Johnson, Jennifer & Trut, Lyudmila & Acland, Gregory (2011) 'Genetics of behavior in the silver fox', *Mammalian genome: official journal of the International Mammalian Genome Society*, 23. 164–77

Nicholls, Henry (5 October 2009) 'My little zebra: The secrets of domestication', *New Scientist*

Norman K., Pellis, S., Barrett, L., & Henzi, S. P. (2015) 'Down but not out: Supine postures as facilitators of play in domestic dogs', *Behavioural Processes*, 110 88–95

Parr L. A., Waller, B. M. (2006) 'Understanding chimpanzee facial expression: insights into the evolution of communication', *Soc Cogn Affect Neurosci*. 2006; 1(3): 221–228

Parsons, K. J., Anders Rigg, A. J. Conith, A. C. Kitchener, S. Harris, & Haoyu Zhu (2020) 'Skull Morphology Diverges between Urban and Rural Populations of Red Foxes

Mirroring Patterns of Domestication and Macroevolution', *Proceedings. Biological Sciences*, 287. 1928
Pilley, John W. (2013) 'Meet the Dog Who Knows 1,000 Words', *Time* magazine, 5 November 2013
Riccucci, Marco (2016) 'Play in bats: general overview, current knowledge and future challenges', *Vespertilio*, 18. 91–97
Rooney, N., Clark, C., & Casey, R. (2016) 'Minimising fear and anxiety in working dogs: a review', *Journal of Veterinary Behavior: Clinical Applications and Research*, 16, 53–64
Rooney, Nicola J., Bradshaw, John W. S., Robinson, Ian H. (2001) 'Do dogs respond to play signals given by humans?', *Animal Behaviour,* Vol.61, Issue 4, pp.715–722
Rooney, N. J., & Bradshaw, J. W. S. (2002) 'An experimental study of the effects of play upon the dog-human relationship', *Applied Animal Behaviour Science*, 75, 161–176
Schenkel, R. (1967) 'Submission: Its Features and Function in the Wolf and Dog', *American Zoologist*, 7 (2) 319–329
Stein, Rob (2004) 'Common Collie or Uberpooch?', *The Washington Post*, 11 June 2004
Terrill, C. (13 March 2012) 'Guarding the Fox House: A famous animal experiment is in peril, after 54 years of work', *Slate*
Trut, Lyudmila N. (1999) 'Early Canid Domestication: The Farm-Fox Experiment: Foxes bred for tamability in a 40-year experiment exhibit remarkable transformations that suggest an interplay between behavioral genetics and development', *American Scientist*, 87 (2): 160–169

The illustration that heads up this chapter is based on a photo by Vegar Abelsnes and is used with permission from Alexandra Horowitz.

Chapter 12

Gantt comes up a couple of times in this chapter and I feel I should stress that, although a believer in the power of dogs for scientific research, Gantt was still very much from the Pavlovian school of scientific enquiry. In other words, he was also partial to putting dogs under 'experimental strain' (read:

abuse and cruelty) as a means to identify new insights into the way their minds worked.

This chapter about love saw me pull on the specialist interests of a number of scientists, mentioned in the opening acknowledgements. However, I am fully aware that there are an enormous number of scientists engaged in research like that specified in this chapter whom I have not called upon. This is something of an editorial choice – my primary aim is to communicate science in a readable way and I became wary in an early draft of this book that a long list of scientists made for a challenging read.

If you are a specialist involved in research projects like those mentioned in this chapter, or any research that helps promote best practice among dog owners, please do look me up on Twitter (@juleslhoward) and I will do all I can to promote your work.

Anderson, S., & Gantt, W. H. (1966) 'The effect of person on cardiac and motor responsivity to shock in dogs', *Conditional Reflex*, 1, 181–189

Berns, Gregory (5 October 2013) 'Dogs are People, Too', *New York Times*

— (2013) *How Dogs Love Us: A Neuroscientist and His Adopted Dog Decode the Canine Brain*. Amazon Publishing

— (2017) *What It's Like to Be a Dog: And Other Adventures in Animal Neuroscience*. Basic Books

Berns G., Brooks A. M., Spivak, M. (2012) 'Functional MRI in Awake Unrestrained Dogs', *PLoS ONE* 7(5): e38027 https://doi.org/10.1371/journal.pone.0038027

— (2015) 'Scent of the familiar: An fMRI study of canine brain responses to familiar and unfamiliar human and dog odors', *Behavioural Processes*, Vol.110, pp.37–46

Bradshaw, J. W. S. & McPherson, J. A. & Casey, Rachel & Larter, S. (2002) 'Aetiology of separation-related behaviour in domestic dogs', *The Veterinary Record*, 151. 43–6

Clive Wynne (2019) 'Dogs' secret superpower: not intelligence, but love', *Science Focus* www.sciencefocus.com/nature/dogs-secret-superpower-not-intelligence-but-love/

David Grimm (16 April 2015) 'How dogs stole our heart', *Science*

Feuerbacher, Erica & Wynne, C. (2011) 'A History of Dogs as Subjects in North American Experimental Psychological Research', *Comparative Cognition & Behavior Reviews*, 6. 10.3819/ccbr.2011.60001

Gantt, W. H. (1973) 'Reminiscences of Pavlov', *Journal of Experimental Analysis of Behavior*, 20, 131–136

Janssens, L., Giemsch, L., Schmitz, R. Street, M., Van Dongen, S., Crombé, P. (2018) 'A new look at an old dog: Bonn-Oberkassel reconsidered', *Journal of Archaeological Science*, Vol.92, pp.126–138

Joseph Stromberg (1 November 2013) 'What fMRI Can Tell Us About the Thoughts and Minds of Dogs', *Smithsonian* magazine

Karin Brulliard (25 September 2019) 'What makes dogs so special and successful? Love', *The Washington Post*

Lazzaroni, M., Range F., Backes, J., Portele K., Scheck K., Marshall-Pescini, S. (2020) 'The Effect of Domestication and Experience on the Social Interaction of Dogs and Wolves With a Human Companion', *Frontiers in Psychology*, Vol.11, p.785

MacLean, Evan L. & Hare, Brian (17 Apr 2015) 'Dogs hijack the human bonding pathway', *Science*, Vol.348, Issue 6232, pp.280–281

Morelle, R. (21 February 2014) 'Dogs' brain scans reveal vocal responses', BBC News

Morey, Darcy F. (2006) 'Burying Key Evidence: The Social Bond between Dogs and People', *Journal of Archaeological Science* 33.2: 158–75

Nagasawa, M., Mitsui, S., En, S., Ohtani, N., Ohta, M., Sakuma, Y., Onaka, T., Mogi, K., Kikusui, T. (2015) 'Oxytocin-gaze positive loop and the coevolution of human-dog bonds', *Science*, Vol.348, Issue 6232, pp.333–336

Ogata, N. (2016) 'Separation anxiety in dogs: What progress has been made in our understanding of the most common behavioral problems in dogs?', *Journal of Veterinary Behavior*, Vol.16, pp.28–35

Sanford, E. M., Burt, E. R. & Meyers-Manor, J. E. (2018) 'Timmy's in the well: Empathy and prosocial helping in dogs', *Learn Behav*, 46, 374–386

Topál, József & Miklósi, Ádám & Csányi, Vilmos & Antal, Dóka (1998) 'Attachment Behavior in Dogs (Canis familiaris): A New Application of Ainsworth's (1969) Strange Situation Test', *Journal of comparative psychology* (Washington, D.C.: 1983), 112. 219–29

Van Bourg, Joshua & Patterson, Jordan & Wynne, Clive (2020) 'Pet dogs (Canis lupus familiaris) release their trapped and distressed owners: Individual variation and evidence of emotional contagion', *PLOS ONE*, 15. e0231742. 10.1371/journal.pone.0231742

Vonholdt, B. M., Pollinger, J. P., Lohmueller, K. E., Han, E., Parker, H. G., Quignon, P., Degenhardt, J. D., Boyko, A. R., Earl, D. A., Auton, A., Reynolds, A., Bryc, K., Brisbin, A., Knowles, J. C., Mosher, D. S., Spady, T. C., Elkahloun, A., Geffen, E., Pilot, M., Jedrzejewski, W., Greco, C., Randi, E., Bannasch, D., Wilton, A., Shearman, J., Musiani, M., Cargill, M., Jones, P. G., Qian, Z., Huang, W., Ding, Z. L., Zhang, Y. P., Bustamante, C. D., Ostrander, E. A., Novembre J., Wayne, R. K. (2010 Apr 8) 'Genome-wide SNP and haplotype analyses reveal a rich history underlying dog domestication', *Nature*, 464(7290): 898–902

VonHoldt, Bridgett M., Emily Shuldiner, Ilana Janowitz Koch, Rebecca Y. Kartzinel, Andrew Hogan, Lauren Brubaker, Shelby Wanser, Daniel Stahler, Clive D. L. Wynne, Elaine A. Ostrander, Janet S. Sinsheimer, and Monique A. R. Udell (2017) 'Structural Variants in Genes Associated with Human Williams-Beuren Syndrome Underlie Stereotypical Hypersociability in Domestic Dogs', *Science Advances*, 3.7

Worrall, S. (9 September 2017) 'Dogs have feelings – here's how we know', *National Geographic*

Wynne, Clive (2019) *Dog Is Love: Why and How Your Dog Loves You*. HMH Books

Yong, Ed (16 April 2015) 'Through This Chemical Loop, Dogs Win Our Hearts', *National Geographic*, 'Not Exactly Rocket Science' blog

Index

A1 (robot) 124–5
Academic Journal of Canine Science 111
advertising, dogs in 95–6
African basenjis 140, 141
African Grey Parrots 173, 265
Albert (child) 64–6
Albert, Ramona 186
'alpha' behaviours, in wolves 97–8, 99, 100–1
alpha dog theory 99–100, 111–12
alpha male, in human culture 98–9
American Kennel Club 96
American Scientist 194
Animal Behavior Enterprises (ABE) 120–2, 131–3
animal intelligence studies 46–52, 66–70
animal rights 36, 241–2
  see also anti-vivisection movement
animal superstitions 116–17, 122, 132
Animal Welfare Act (1966) 263
animats 123–7
anthropomorphism 19, 48, 172
anthropopsychism 52
anthrozoology 229
anti-vivisection movement 44, 55–8, 77–89, 255
  brown dog affair 80, 82–3, 84–8, 255
  Cruelty to Animals Act (1876) 81–2
  The Shambles of Science 79–81, 82–3
  in USA 152, 263
ants 69, 167
apes
  bonobos 105, 167, 182
  chimpanzees 50, 165–8, 178–80, 182, 190, 199, 221
  cognitive research 164–8, 178–80, 181–2
  compared to humans 188–9
  deception studies 178–80
  gorillas 167, 182
  mirror self-recognition (MSR) test 164–8
  orangutans 164–5, 167
  play 199
  theory of mind studies 179–80, 181–2
  yawn contagion 221
archaeological discoveries 225–6
Aristotle 27
artificial selection 195
assistance dogs 122, 143
attachment 186, 221–3
attention-getting behaviours 202–4
aversive conditioning 147–9

baboons 105, 221
Baby Tender 119
bacteriology 77
Bar Harbor study 139–46
Basel Zoo 100
Bates, Alan W. H. 33
bats 157–9, 196
Battersea Dogs' Home 43–4
Battersea, London 84–8, 255
Bayliss, William 83
beagles 140
Beck, Aaron 149
behavioural genetics 137–46
Behaviourism 64–6, 70, 89, 113–22, 130–4, 152
Bekoff, Marc 19, 169–70, 177, 196–7
Belyaev, Dmitry 193–6
Bentham, Jeremy 73
Bergamasco Sheepdog 94
Bergson, Henri 74
Bernard, Claude 57–8
Berns, Gregory 232–5, 241, 243
Berzelius, Jöns Jacob 137
Betsy (dog) 211, 232
biases 19–20
Biff (dog) 117, 231–2, 239–40
Bloom, Paul 210
bonobos 105, 167, 182
border collies 104, 205–13

Bradshaw, John 100–1, 111, 177, 201, 222–3
brain 63, 73–6
caudate nucleus 234, 235
cerebellum 197
entorhinal cortex 134
fMRI research 163, 233–6
hippocampus 134
mental maps 134–5
place cells 134
play and 197
sense of smell and 162
size in dogs 177
somatosensory cortex 163
see also cognitive sciences
breeding, selective 94, 193–6, 242
breeds, dog 93–6
Breland, Keller 120–2, 131–3, 135
Breland, Marian 120–2, 131–3, 135
Brillat-Savarin, Jean Anthelme 162
British Medical Association 87
British Medical Journal 56, 255
British Union for the Abolition of Vivisection (BUVA) 55–6, 88, 255
brown dog affair 80, 82–3, 84–8, 255
Budiansky, Stephen 107
Buffon, Georges-Louis Leclerc, Comte de 27–8
Burghardt, Gordon M. 197
Buzan, Deborah Skinner 119

Callie (dog) 233
Cambridge Declaration on Consciousness 265
canine distemper 226
captive wolves 100
cardiovascular conditioning 217
caudate nucleus 234, 235
cerebellum 197
Chambers, Robert 30
Chaser (dog) 22, 211–13, 232
children
Baby Tender 119
conditioning experiments 64–6
depression treatment 150–1
'Little Albert' 64–6
rabies vaccine trials 55
Sally-Anne test 180–1
chimpanzees 50, 165–8, 178–80, 182, 190, 199, 221
Chittka, Lars 69
citizen science 267
classical conditioning 59–66, 70–1, 217
Clever Hans (horse) 208–9
clicker training 121–2
Cobb, Frances Power 44, 55–6
Cobb, Matthew 74
cocker spaniels 140, 141
cockroaches 68
'cocktail party' type of personality 227
Cognitive Behavioural Therapy (CBT) 147, 149
cognitive sciences 123–35, 152
animats 123–7
ape studies 164–8, 178–80, 181–2
consciousness 159–60, 164–74, 175–6, 265
deception studies 178–80
feedback loops 127–8
innate behaviours 131–4
mental maps 134–5
mirror self-recognition (MSR) test 164–70
number seven 129
place cells 134
Sally-Anne test 180–1
smell-based self-recognition tests 169–71
theory of mind 179–82, 200–5, 212
wolf studies 185–6, 190
see also dog cognition research
Coleridge, Stephen 83
commensalism 14
communication, social 103–8
conditioning
aversive 147–9
cardiovascular 217
classical 59–66, 70–1, 217
operant 70, 112, 113–22, 180
unconditioning 66
consciousness 52, 68, 73–4, 159–60, 164–74, 265
consent forms 243
Coppinger, Lorna 39–40
Coppinger, Raymond 39–40
Coren, Stanley 111

Covid-19 pandemic 17, 119
Crick, Francis 175, 176
crickets 126–7
critical period of development 141–2, 146
cruelty and mistreatment of dogs
  aversive conditioning 147–9
  electric shock experiments 147–9, 261–3
  genetic inbreeding 242
  medical research 55–8, 77–89, 152, 255, 263
  Pavlov's studies 60–1, 70–1
  puppy farms 17, 242
  street dogs 41–3
  training methods 99–100, 109–10, 111
  vivisection 55–8, 77–89, 152, 255, 263
Cruelty to Animals Act (1876) 81–2
Csányi, Vilmos 183–5, 222
cybernetics 127

Daily News 83
Darwin, Charles 28–9, 30–2, 45–6, 48, 50–1, 53, 78, 130, 157, 164–5, 222, 236
Dawkins, Richard 171
de la Ramée, Marie Louise 54
deception studies 178–80
Dennett, Daniel 160
depression 148, 150–1
Descartes, René 75
Despard, Charlotte 85
Dewsbury, Donald A. 145
Dick, Philip K. 171
Dickens, Charles 33–5, 43–4, 80–1
digestive system research 60–2
digging behaviour 104–5
distress-anxiety theory of destructive behaviour 186, 222
'Do-As-I-Do' requests 212
dog breeds 93–6
  behavioural differences between 139–41
  hybrid vigour 139
dog cognition research 21–2, 182–92
  citizen-science projects 267
  decline in 176–8
  'Do-As-I-Do' requests 212
  Family Dog Project 185–8, 190–2, 222, 235
  fast-mapping 210
  many-to-one mapping 211
  mirror self-recognition (MSR) test 168–9
  object retrieval studies 206–8, 209–13
  pointing gesture studies 188–92, 230
  smell-based self-recognition tests 169–71
  theory of mind studies 200–5, 212
dog experiments
  aversive conditioning 147–9
  behavioural genetics research 139–46
  brain research 75, 163, 232–6
  classical conditioning 59–64, 70–1
  digestive system research 60–2
  early animal intelligence studies 47–52, 66–8
  electric shock experiments 147–9, 261–3
  ethical treatment 142, 152–3, 243
  fMRI research 163, 233–6
  heart rate studies 217–18
  'Kaspar Hauser' study 144
  learned helplessness 147–9, 261–3
  medical research 55–8, 77–89, 152, 255, 263
  oxytocin studies 219–21, 223
  Pavlov's studies 59–64, 70–1
  personality development 139–46
  puzzle-box experiments 66–8, 70
  reflex actions 75–6
  vivisection 55–8, 77–89, 152, 255, 263
  see also dog cognition research
Dog Owners' Protection Association (DOPA) 54
dog play behaviours 197–205
  attention-getting behaviours 202–4
  object retrieval studies 206–8, 209–13
  play-bows 198–9
  play signals 202, 203
  rolling over 199
  theory of mind studies 200–5
  tug-of-war games 200–1
dog training 97, 109–12, 115
  alpha dog theory 99–100, 111–12
  assistance dogs 122, 143
  bridging stimuli 121–2
  clicker training 121–2
  cruel methods 99–100, 109–10, 111

innate behaviours and 133–4
play and 200
positive reinforcement 112, 121–2
dog-whippers 42
Dognition project 267
dogs
in advertising 95–6
alpha dog theory 99–100, 111–12
behavioural differences between breeds 139–41
brain size 177
breed diversity 93–6
digging behaviour 104–5
domestication 14–15, 32, 38
eye-gazing 219–21
facial expressions 107–8
as film stars 95, 97, 110
grief at deaths of 240–1
hybrid vigour 139
hyper-sociability 227–8
innate behaviours 133–4
legal rights 36, 241–2
names of Pavlov's 71, 253
personality development 139–46
pointing behaviour 104
population growth 16–17, 95, 96, 247
puppy-dog eyes 107–8
registrations 95, 96
retrieval of objects by name 206–8, 209–13
scratching reflexes 75–6
self-recognition 168–71
sense of smell 160–4
separation distress 186, 222–3
sleeping behaviour 106–7
social attachment 186, 221–3
social behaviours 41, 103–8
social reflex 217–18
street dogs 36–44, 58
submission behaviours 103, 105–6, 108–9
superstitious behaviours 117
tail-wagging 105
thermal radiation detection 163
urine marking 104
village dogs 37, 38–41
yawn contagion 221
see also dog play behaviours; puppies
Dogs for Defence (DFD) programme 110
Dogs Trust 17
Dóka, Antal 185
dolphins 50, 158, 167, 196
domestication
dogs 14–15, 32, 38
red foxes 195
silver foxes 193–6
dominance behaviour, wolves 97–8, 99, 100–1
dopamine 118, 234
Du Bois, William 69
Duval, Mathias-Marie 75
dyspepsia 61

echolocation 158–9
electric shocks
in dog training 99, 100
experiments using 147–9, 261–3
elephants 167, 196
'Emancipation Day' 58
Engber, Daniel 263
entorhinal cortex 134
ethical treatment 142, 152–3, 243
ethology 129–30, 142, 183–92
eugenics 146
Evans, Mark 242
evolution 26–32
eye contact 219–21

facial expressions, dogs 107–8
faeces, human 38–9
Family Dog Project 185–8, 190–2, 222, 235
Farage, Nigel 99
fast-mapping 210
fear response 65–6
feedback loops 127–8
feminist movement 80
feral dogs 37–8
films
consciousness themes 171–2
film star dogs 95, 97, 110
Finnish Lapphund 94
First World War 109–10
Fischer, Julia 208, 209–11
Fitzgerald, Percy 34
flank-scratch reflex 76

Fleury, Gabi 112
Flip (dog) 21–2, 182–4, 232
foxes
  red 195
  silver 193–6
free will 114
Frisch, Karl von 129, 130
Fuller, John Langworthy 137–46, 151–2, 222
functional magnetic resonance imaging (fMRI) 163, 233–6

Galambos, Robert 158
Galloway, John 255
Gallup, Gordon G., Jnr. 165–8
Gålmark, Lisa 80
Galpayage, Samadi 69, 252–3
Gannt, W. Horsley 217–18
gastric juices 60–2
Gatti, Roberto Cazzolla 170
genetic inbreeding 242
Genetics and the Social Behavior of the Dog 142–3, 146
German shepherds 95, 96
gestures, human 188–92, 196, 230
Giemsch, Liane 226
golden retrievers 96
gorillas 167, 182
Gray, Jay 111
grey wolves *see* wolves
Greyfriars Bobby (dog) 33
grief, at deaths of dogs 240–1
Griffin, Don 157–9
Griffin, Donald 172–3
Griffin (parrot) 173
Grimm, David 242
GTF2I/GTF2IRD1 genes 227–8
guide dogs 122, 143

hand gestures, human 188–92, 196, 230
Hardy, Thomas 80–1
Hare, Brian 189–90, 191, 220, 230, 267
Harvey, Naomi 112
heart rate studies 217–18
Herzog, Hal 95, 96, 263
hippocampus 134
Hollywood dogs 95, 97, 110
Homans, John 53
hormones
  dopamine 118, 234
  oxytocin 218–21, 223
Horowitz, Alexandra 19, 162, 170, 198–9, 201–5
Howell, Philip 55
Huizinga, Johan 193
human–dog burials 225–6
human evolution timeline 9–11
hunting behaviour, wolves 101–2
huskies 96
Huxley, Thomas Henry 25, 30, 47, 74
hybrid vigour 139
hyper-social dogs 227–8

If Dogs Could Talk 183–4
imprinting 130, 142
innate behaviours 131–4

Jackson Laboratory, Bar Harbor 139–46
Janssens, Luc 225–6
Jeffery, Kate 134
Jenny (orangutan) 164–5, 167
Jones, Mary Cover 66, 254
Jones, Steve 29

Kaminski, Juliane 107, 208, 209–11, 213
kangaroos 203
Kano, Fumihiro 182
'Kaspar Hauser' study 144
kennel clubs 95, 96
Khan, Niki 145–6
Kikusui, Takefumi 219–20
kin selection 102
King's College London 79
Kipling, Rudyard 80–1
Kirk, Robert 56
Koch, Christof 175–6
Koehler, Bill 111
Koehler, William 115
Koko (gorilla) 167
Konnikova, Maria 150
Kruse, Marian *see* Breland, Marian

laboratory dogs
  aversive conditioning 147–9
  brain research 75
  classical conditioning 59–64, 70–1
  digestive system research 60–2

electric shock experiments 147–9, 261–3
ethical treatment 142, 152–3, 243
learned helplessness 147–9, 261–3
medical research 55–8, 77–89, 152, 255, 263
Pavlov's studies 59–64, 70–1
reflex actions 75–6
vivisection 55–8, 77–89, 152, 255, 263
see also dog experiments
Lady Jane (orangutan) 164–5, 167
Lao Tzu 113
lap-dog breeds 96
Law of Effect 70
lawsuits 242
learned helplessness 147–51, 261–3
learning curves 67–8
learning, trial-and-error 47–8, 66–8
legal rights 36, 241–2
levator anguli oculi medialis (LAOM) muscle 107
Lévi-Strauss, Claude 123
Life magazine 263
Lind-af-Hageby, Emilie Augusta Louise 'Lizzy' 77–80, 82–3, 88, 236
Lind, Lizzy 44
'Little Albert' 64–6
London Battersea 84
London School of Medicine for Women 79
Lorenz, Konrad 108–9, 129, 130, 142
love, of dogs for humans 215–16, 228–30
brain research 232–6
genetic evidence 226–8
heart rate studies 217–18
oxytocin studies 218–21, 223
separation distress 186, 222–3
Lubbock, John 48–50, 52, 222

Magendie, François 56–7
Maier, Steven 148, 149
many-to-one mapping 211
marketing, dogs in 95–6
Marla (dog) 22, 228, 232
Martin, Marie Françoise 57–8
maths-teaching machine 118–19
Mayr, Ernst 28
Mech, Dave 99
medical research 55–8, 77–89, 152, 255, 263
medical student protests 86–7
Meister, Joseph 55
'mental evolution' 46–8
mental maps 134–5
Miessner, B. F. 124
Miklósi, Ádám 184–8, 190–2, 222, 267
military dogs 109–11
Mill, John Stuart 30
Millan, Cesar 99–100, 111
Miller, George Armitage 129, 130–1, 135
mirror self-recognition (MSR) test 164–70
'The Misbehavior of Organisms' 133
Morgan, Conwy Lloyd 46–8, 50, 52, 73–4
Morgan's Canon 48, 50, 209
Most, Konrad 109–10
Musi, Vince 211
mutual eye-gazing 219–21
mutualism 14
muzzles 54–5, 58
myelin sheath 75

Nagel, Thomas 158–60, 164, 171, 173
National Anti-Vivisection Society 83, 88, 255
National Geographic 211, 220–1, 226
natural selection 29, 73–4
Nature 255
Nazi party 109
negative reinforcement 115
neonatal period 141
nervous system 56, 75–6
Neumann, John von 128
neurotransmitters 118, 234
New Scientist 191
New York Daily Times 42
New York Times 85–6, 241
New Yorker 150
Newton, Isaac 59
Noel, Charles 85
Nordic Anti-Vivisection Society 79
Norwegian Buhund 94
Norwegian Lundehund 94
number seven 129

Oberkassel Dog 225–6
object retrieval studies 206–8, 209–13
O'Keefe, John 134
On the Origin of Species 29, 30–2, 45–6, 78
operant conditioning 70, 112, 113–22, 180
orangutans 164–5, 167
Oreo (dog) 21, 189–90, 232
owned-restricted dogs 37
Oxford University Museum 30
oxytocin 218–21, 223
Oz (dog) 230–1

Pal, Sunil Kumar 40–1
parasitism 14
pariah dogs of India 39, 40–1
parrots 173, 265
Parsons, Kevin 195
Pasteur, Louis 55, 77
pasteurisation 77
Pavlov, Ivan 59–64, 70–1, 110, 114, 217, 253
PDSA 16, 122, 143, 145
Pekinese 96
Penn Resiliency project 151
Pepper (dog) 263
Pepperberg, Irene 173, 263
personality development in dogs 139–46
pet industries 96
Peter (child) 66
Pfungst, Oskar 209
phosphine 224
pigeon-guided missile systems 120
pigeons 115, 116, 167
Pilley, John W. 212
place cells 134
Planet of the Apes 172
play 193
  defined 196–7
  functions of 197
  see also dog play behaviours
Plott hound 94
pointing behaviour 104
pointing gestures 188–92, 196, 230
poodles 96, 106–7
population growth, dogs 16–17, 95, 96, 247
positive reinforcement 112, 115, 121–2
postal workers 117
prairie voles 219
Preece, Rod 89
Premack, David 178–80
problem-solving *see* puzzle-box experiments
programmed instruction 119
Project Pelican 120
pugs 96
Punch 95
puppies 138
  personality development 139–46
  prices 17
  sensitive period of development 141–2, 146
  separation distress 186, 222–3
  socialisation 141, 143, 145
puppy-dog eyes 107–8
puppy farms 17, 242
puppy parties 143
puzzle-box experiments 66–8, 70, 114–15, 116, 119

Question of Animal Awareness, The 172–3

rabbits, fear of 66
rabies 41, 42–3, 53–5, 58
rabies vaccine 55
raccoons 121, 132
Ranvier, Louis-Antoine 75
rats 115, 134, 196
  fear of 65–6
reciprocal innervation 76
red foxes 195
reductionism 114, 135
reflex actions 75–6
  see also classical conditioning
reinforcement 110, 115–17, 180
  negative 115
  positive 112, 115, 121–2
retrievers 96, 104, 189–90, 200–1
Rico (dog) 22, 205–8, 209–11, 213, 232
Rin Tin Tin (dog) 95
Robinson, Lauren 111
robots 123–7, 171
Romanes, George 50–2, 73, 222
Rooney, Nicola 200–1
Roux, Emile 55
Royer, Clémence 30
RSPCA 223, 242

salivation 62–3
Sally-Anne test 180–1
Schartau, Leisa Katherine 79–80, 82–3, 88
Schenkel, Rudolph 97–8, 99, 100, 101, 108
Science 191, 210, 220
Science Focus 232
Scott, John Paul 137–46, 151–2, 222
Scott, Sophie 235–6
Scottish terriers 140
scratching reflexes 75–6
seals 203
Second World War 109, 110
Secord, James A. 30
selection
  artificial 195
  kin 102
  natural 29, 73–4
selective breeding 94, 193–6, 242
Seleno (robot) 123–4, 127
self-mutilation 222
self-recognition 164–71
Seligman, Martin 147–52, 261–3
sensitive period of development 141–2, 146
separation distress 186, 222–3
Shambles of Science, The 79–81, 82–3
Shaw, George Bernard 85
Sherrington, Charles Scott 76
shock-collars 99, 100
shuttle-boxes 148, 149, 261–3
silver foxes 193–6
Skinner, B. F. 113–22, 130, 133, 134
Skinner Box 114–15, 116, 119
sleeping behaviour 106–7
Sloughi 94
smell-based self-recognition tests 169–71
smell, sense of 160–4
Sniff Test of Self-Recognition (STSR) 170
Sobel, Dava 115
social attachment 186, 221–3
social behaviours 41, 103–8
social hunting, wolves 101–2
social reflex 217–18
socialisation 141, 143, 145
socialisation period 141, 230
Society for the Prevention of Hydrophobia and the Reform of the Dog Laws (SPH) 54
Society for the Protection of Animals Liable to Vivisection (SPALV) 55–6
Solaris (film) 172
Solicitors Act (1843) 78
somatosensory cortex 163
Spalding, Douglas 142
spiders 68
Spitz, Carl 110–11
Starling, Ernest 83
statue of brown dog 84–8, 255
Stevenson, Robert Louis 215, 237
Stillwell, Victoria 111
stone tools 10
stray dogs *see* village dogs
street dogs 36–44, 58
Strongheart (dog) 95
structured socialisation 143
submission behaviours 103, 105–6, 108–9
suffragettes 86
superstitious behaviours 116–17, 122, 132
Sutherland, Stuart 176

tail-wagging 105
Taylor, Shelley 218
Tealby, Mary 43
Teo, Celeste 241
terriers 33, 96, 140
territorial defence, wolves 102–3
theory of mind 179–82, 200–5, 212
thermal radiation detection 163
Thorndike, Edward 66–70
Time Magazine 212
Tinbergen, Nikolaas 129, 130
Todes, Daniel T. 61, 63
toilet-roll timeline 9–11
Tomasello, Michael 188–90
Tony (dog) 47–8
Topál, József 185
Townsend, Emma 32
training 97, 109–12, 115, 120–2
  alpha dog theory 99–100, 111–12
  assistance dogs 122, 143
  bridging stimuli 121–2

clicker training 121–2
cruel methods 99–100, 109–10, 111
innate behaviours and 133–4
play and 200
positive reinforcement 112, 121–2
transition period 141
trial-and-error learning 47–8, 66–8
Trump, Donald 99
Trut, Lyudmila 194–6
tug-of-war games 200–1
Tully, Tim 253
turbinates 161
Turing, Alan 128
Turner, Charles H. 68–9, 130, 252–3
Twain, Mark 53, 93, 108

Uexküll, Jakob Johann von 159
umveldt 159
unconditioning 66
University College London 79, 83, 255
urine marking 104

vaccines 55, 77
Van (dog) 49–50
Vänskä, Annamari 96
Venus 224–5
Victoria, Queen 81
village dogs 37, 38–41
see also street dogs
vivisection 55–8, 77–89, 152, 255, 263
Vogt, Carl 31
Vogue magazine 96
voles 203, 219
vonHoldt, Bridgett 227–8

wallabies 196
Wallace, Alfred Russel 28, 74
Washington Post 210
wasps 69
Watson, John B. 64–6
Wells, H. G. 114
Wensley, Sean 143, 145
Wetten das? (TV programme) 205–8
'What Is It Like to Be a Bat?' 159–60
Wiener, Norbert 127–8, 135
Wilberforce, Samuel 30
Williams-Beuren syndrome 227
wolves 38
captive 100
cognitive research 185–6, 190
dominance behaviour 97–8, 99, 100–1
eye contact 220
hunting behaviour 101–2
lone 102–3
oxytocin 223
packs 102–3
sleeping behaviour 106–7
sociability 228
social behaviours 103–8
submission behaviours 103, 105–6
territorial defence 102–3
Woodard, Colin 192
Woodruff, Guy 178–80
Woods, Abigail 54
World League Against Vivisection 84
Wynne, Clive 228–30, 232

Xephos (dog) 229–30, 232

yawn contagion 221
yellow snow test 169–70
Yin, Sophia 133
Young, Larry 220–1

Zimen, Erik 106–7

JAN 1 9 2023